# Sweet in Tooth and Claw

Kristin Ohlson is a writer based in Portland, Oregon. Her work has appeared in *The New York Times*, *The Christian Science Monitor*, *Salon*, *Discover*, and elsewhere. She is co-author of the bestselling *Kabul Beauty School* and author of *The Soil Will Save Us: how scientists, farmers, and foodies are healing the soil to save the planet.*

Kristin Ohlson

# Sweet in Tooth and Claw

*nature is more cooperative than we think*

SCRIBE
*Melbourne • London*

Scribe Publications
2 John St, Clerkenwell, London, WC1N 2ES, United Kingdom
18–20 Edward St, Brunswick, Victoria 3056, Australia

Published by Scribe 2022

Typeset in Bembo 13/20 pt by the publishers

Printed and bound in the UK by CPI Group (UK) Ltd,
Croydon CR0 4YY

Scribe Publications is committed to the sustainable use of natural
resources and the use of paper products made responsibly from those
resources.

978 1 911617 34 1 (UK edition)
978 1 925713 16 9 (Australian edition)
978 1 922586 59 9 (ebook)

Catalogue records for this book are available from the National Library
of Australia and the British Library.


scribepublications.co.uk
scribepublications.com.au

For Jamie Rose, my sweetest.

The whole point of our evolution, it seems to me, is for us to find a way to fit back into the world as it is, rather than try to remake the world to fit us.

*Brian Doyle*

When we love the earth, we are able to love ourselves more fully.

*bell hooks*

May all living things find sweetness and ease.

*my yoga teacher Jeannie Songer's favourite translation of Loka Samasta Sukhino Bhavantu, a chant used in meditation*

# Contents

# Preface

Years ago, I joined a group of women and a few men at an art gallery in the Murray Hill neighbourhood of Cleveland, Ohio. It was supposed to be an evening of journalling and shared feelings and uplifting conversation — stuff that makes me a little skittish, then more than now — and I can't even imagine now what convinced me to go. I remember so little about the event — was the Lego-like brick street outside slick with ice, as it so often was? Did I still have young children waiting for my return at home? But one trick of the convener stays with me still. We were all sitting on the floor, knee to knee, and she asked us to look around the room and make a mental list of all the blue things in the room. There were quite a few of them, and I quickly jotted them down in my brain. Then we closed our eyes, and one by one, the convener asked us an unexpected question. Name one of the yellow things in the room! As I recall, no one — and certainly not I — could remember anything yellow because we were all so intently focused on blue. The yellows had faded into the background along with the greens and purples and reds, erased by our inattention.

The exercise buttressed one of the themes the convener would pursue that day, that what we decide to focus on not only informs our view of the world, but will also guide our

1

path through it. I found that idea valuable as my life went on its twists and turns, especially during times of fear or despair. During the recent pandemic, for instance, our walks with the dogs offered many small distracting wonders if I remembered to look for them: moss growing in old lettering on the sidewalk, clouds of excitable bushtits settling into an overgrown shrub, bark peeling like ringlets off a birch tree. All the couples walking hand in hand around the neighbourhood, keeping a respectful six-foot distance from everyone else. Neighbours having tiny, candlelit parties on their lawns throughout the summer and fall. The young musicians who brought a bass and a violin to the small park across the street from my house and played for an hour. People who went into a frenzy of giving, as much as they were able, to others harder hit by the disaster. As the wonderful Fred Rogers (Mr Rogers) said on the first anniversary of 9/11, 'Look for the helpers.' There were so many helpers.

I'm naturally drawn to optimism, which is a gift from my sweet father. I actually worried that I might just be soft-headed until I read this quote from activist and professor Angela Davis: 'I don't think we have any alternative other than remaining optimistic. Optimism is an absolute necessity, even if it's only optimism of the will … and pessimism of the intellect.' But it's hard to hang on to optimism. Like others — probably you — I panic at the growing, undeniable evidence of humanity's damage to the natural world around us, and fear we'll never get our shit together to do anything about it as our politics and cultures continue to clash in the nastiest of ways. When I wrote my previous book, *The Soil Will Save Us*, I discovered a wellspring of optimism as I met farmers, ranchers, scientists, and others figuring out how to restore damaged agricultural landscapes. But if the world is characterised by greed and grasping and selfishness, as so many people believe, would the growing numbers of ordinary ecological heroes be enough?

Then I heard Canadian forest ecologist Suzanne Simard speak at the 2015 Urban Soil conference in Los Angeles. For the last thirty years, she's been uncovering the hidden cooperation among trees and other living things in the forest. In the process of writing *The Soil Will Save Us*, I was thrilled to learn about the life-giving partnership between plants and soil microorganisms. Really, it was the greatest of revelations to find out that plants don't just suck nutrients from the soil and leave it as barren of goodness as a Twinkie, but are engaged in a constant give and take with the billions of tiny organisms there. At the conference, Simard talked about this kind of fertile partnership spread out across the forest landscape, powered by a vast underground skein of fungi. I almost levitated from my seat with excitement.

I drove from Portland to Vancouver that year to talk to Simard and her students, but it took me a few years to find other researchers and landscapes yielding similar insights. As they accrued, I felt as if I was really on to something worth writing about. Most of us have forgotten much of what we learned in science class, but a few concepts persevere —

and 'survival of the fittest' is certainly one of them. Charles Darwin posited that the species we see around us today are the winners of a challenge that's been going on for nearly four billion years; all living things are the culmination of changes passed on from ancient forebears that made them more successful at harvesting resources, avoiding predation and other dangers, and reproducing. Other thinkers pounced on Darwin's findings and enshrined the concept of competition as biology's brutal architect. The idea that competition rules has been lodged in our collective brains ever since. Even if people reject the theory of evolution or can't quite recall how it works, they still think of nature as 'red in tooth and claw,' as the poet Tennyson wailed — a vicious and never-ending battle of survival for meagre resources. Even many scientists don't grasp how pervasive cooperative interactions are in nature. 'Today's ecologists grew up with the paradigm of organisms primarily competing with each other,' biologist Richard Karban told me. 'A lot of ecologists are surprised by how much cooperation exists among plants and other organisms. They're not looking for it in their research.' Consequently, we seem to have developed a zero-sum-game view of nature, suggesting that whatever we take — we humans, or those ravens, cypresses, invasive garlic mustard, or any living thing — comes at the expense of other living things and the overall shared environment. As we humans keep growing in number, this view suggests, it must regrettably follow that the rest of the world will suffer.

But what if we're applying Darwin's insights wrongly to the world and thus missing the generosity and cooperation that exist in the natural world? That's what Simard's research suggested to me. And if we are missing the generosity and cooperation in the greater world, we are likely also missing these harmonious connections in ourselves. Because we are part of nature, of course. We exist because of complex, vibrant, creative relationships with the rest of nature and are as much a part of it as the raccoon lounging in the tree near my front door or the grasses growing along the highway. How might our behaviour change if we understood the extent to which cooperation within and among species undergirds the natural world and makes it thrive? If we looked for that cooperation, just as I was instructed to look for the blue things in that gallery? Could we begin to see ourselves as partners and helpers, part of a greater fabric of giving, instead of exploiters and colonisers and wreckers?

It seems to me that the best and highest use of science these days is to figure out how nature works — as many of the brilliant scientists I interviewed in this book are doing — and then to help humans change our behaviour so that we can roll back the damage we've already done and avoid further damage. So that we encourage and bolster the world's hunger to thrive. And not just because that would benefit us, although it certainly would, but because other life forms have as much right to flourish as we do and don't exist for our use. Many scientists are learning from the ways in which older cultures understood their place within nature and how they balanced human needs with the needs of the other living things. I'm convinced that if we can learn to respect, not ravage, the rest of nature, we'll also become more generous and nurturing with each other. 'What you do to the people, you do to the land,' says Gopal Dayaneni, an activist with the Movement Generation Justice and Ecology Project. 'And what you do to the land, you do to the people. This is a common concept across many Indigenous and land-based traditions.'

The person who might best articulate this respectful bond with the rest of nature — at least, I'm smitten with her writings and recordings — is the Native American botanist and writer Robin Wall Kimmerer. Humans must take from nature to survive, she says, but we have to make sure it's an honourable harvest. Never take the first plant or animal or the last. Ask permission: the world is generous and creative but sometimes the answer is no, and science is a powerful tool for understanding that limit. Do as little harm as possible. Practise gratitude. Share. And reciprocate — we need to learn how to give back.

We humans are engaged in so many harvests, and they are often not honourable. When we take fish from the ocean or grow tomatoes in a field; when we seize land from the prairies or forests to make a home or a city; when we divert water for our urban areas or for agriculture; when we take the labour or the confidence of other humans ... all are an opportunity for honour. Parts of this book look at the cooperation and mutually beneficial connections that hold together the creatures and ecosystems in and around us; I find that to be thrilling science. But for me, the most thrilling parts of the book look at how people are acting on new understandings of what the rest of nature needs from us. They are deciding to be partners with the living world, partners with each other in this mission, and showing that bleakness does not have to be our shared fate.

## Chapter 1

# An Underground Tapestry
# of Give and Take

As we leave the gravel road and step into the forest, it seems more like a dream of a forest than a real place. Or maybe it's real but we aren't. We can't hear our footsteps because the forest floor is deep and muffled with fallen needles and branches and other debris, all silently acquiescing to the hidden hungers of microscopic life. We can't hear our voices — at least, I can't hear their voices when they disappear from view, and I startle at the sudden and complete silence. I walk faster to catch up and not be the pain in the ass who interferes with their work by getting lost. Because I talked myself onto this research trip and offered to help set up the site, and I'd hate to make them regret saying yes.

The forest floor is so soft and springy that I feel a little seasick. I follow a will-o'-the-wisp of laughter to a cluster of young trees, and then the pitch of the land changes and I'm plunging downhill. I don't worry about falling because all the hard edges of the forest seem so padded with either greenery or rot that I don't think a tumble would hurt. I pass great humped shapes among the trees — boulders or maybe fallen trunks, although they look like whales surfacing, velveted in brown and green. I finally catch sight of University of British Columbia ecologist Suzanne Simard's blonde hair, which might be the brightest thing in the entire Malcolm Knapp Research Forest, and rush to catch up.

We're still surrounded by living trees, but here and there are ageing stumps that are all around the same height. Some still have patches of bark, most are slicked with white and green lichen, and some are encrusted with charcoal. A few have a little fright wig of splinters on top. All have a gaping hole a few feet from the ground that's about as big as my two fists put together — a hole that's flat on the bottom and frowning down on top. The stumps look like many things — small statues of the Easter Island mo'ai, eyeless tiki gods, the unhappy ghosts of trees, all shrieking soundlessly about the massacre they suffered and the upcoming harvest scheduled for this very spot.

'Why do they have such sad faces?' I ask Simard and Jean Roach, her long-time friend and forestry colleague.

'This forest was harvested by hand back in the 1940s,' Simard tells me. 'The men made those holes so they could stick in a piece of wood called a springboard. They'd stand on that to make the cut higher up the stem, away from the curvature of the base.'

'Eighty years ago!' I say. 'I'm amazed these stumps are still here after all that time.'

She looks around. 'This was an old-growth cedar forest. Now it's cedar mixed with hemlock and Douglas fir, but if we let it go for 500 years the cedar would retake it.'

She and Roach continue tramping down the slope, following their GPS to the centre of the research site. I take a few pictures of the sad-sack stumps before I follow them, feeling a little silly about my rush to anthropomorphise. But screw it, I decide: if Canada's timber industry didn't want the odd wanderer to read these stumps as an indictment of humanity's harsh hand in the forest, they should have whittled those holes into smiley faces.

After I heard Simard speak at the Urban Soil conference in 2015, she seemed to be in the news more and more. The science media were paying lots of attention, and there were several of her TED Talk videos bobbing around on the internet, as well as an audio interview on Radiolab. By the time I was ready to visit her for a second time in 2017, I was afraid she was so inundated with requests from people like me that she wouldn't have the time or the bandwidth for my questions. And that seemed to be the case for a while, as my emails and calls went unanswered. When I finally heard from her, she was apologetic: family complications and weeks spent deep in the forest had intervened. She invited me to spend a few days with her in British Columbia in the fall, helping her and Roach set up plots for a huge, multisite study she had launched called the Mother Tree Project. And if I wanted to get a better idea of what the project was about, I could drive up the day before the work began in Malcolm Knapp and hear her explain it to a group of commercial foresters near Merritt, British Columbia.

I arrived in Merritt to meet up with Simard and her postdoc, Teresa (Sm'hayetsk) Ryan, whose research focuses on increasing the understanding of how the Indigenous people of North America cared for the land. I had been wondering if Simard's work might seem either threatening or irrelevant to men and women who make a living cutting down trees, but the foresters were friendly as we trooped into the woods. There, Simard and Ryan offered a quick summary of the Mother Tree Project, Simard's biggest research effort yet, with support from several universities, First Nations, and government. It focuses Simard's lifetime of research on the question of how to harvest trees in a way that allows forests to regenerate more quickly — an urgent question during this period of climate change, in which wet forests are becoming drier and forests are stressed as never before. Many people agree that the conventional practice of clear-cutting — logging all trees in a section of forest — and replanting them with seedlings grown in sterile, nursery soil needs urgent reassessment: clear-cutting is not only a body blow to the forest, but also burps a big pulse of carbon formerly stored in forest soil into the air. Forest soil loses half its carbon in a clear-cut. According to Greenpeace, the carbon emissions from deforestation comprise a fifth of global emissions, higher than those from transportation. However, it's unclear which logging practices are better for long-term forest and climate health. The Mother Tree Project hopes to answer that.

In six forests dominated by Douglas firs across a climate gradient, the project will test four different harvesting strategies: conventional clear-cutting, single-tree retention (in which single trees will be spared every twenty to twenty-five yards, with preference for the larger and older Douglas firs that Simard calls Mother Trees), retention of 30 per cent of the forest in patches centred on Mother Trees, and retention of 60 per cent of the forest in patches centred on Mother Trees but with some of the smaller trees within the patch removed for harvest. In each site, they will also leave an unharvested plot as a control. In each of the harvested areas, they will test various replanting and seeding treatments. In the coming years and decades, researchers will return to these sites to assess forest regeneration, productivity, soil-carbon levels, and resilience.

The foresters had many questions, but none expressed any of the scepticism I anticipated about Simard's basic premise that trees are engaged in a daily dialogue more vigorous than that among my neighbours. One forester asked, 'Why do you call them Mother Trees?' and Simard answered, 'Because they nurture their young and they're big and old!' That got a laugh, presumably because several of the women there, including Simard, are mothers but are not big or old.

Simard had an easy time engaging with the foresters, at least in part because her French-Canadian family has a long connection to the hard work of harvesting trees. Her great-grandfather and his brothers, her grandfather and his brothers, and her father and uncles were all loggers. She adored her grandfather, who lived at the edge of a cedar-hemlock forest and cut small cedars for later use as telephone poles, removing them from the forest by horse and floating them across a lake to be milled. 'He logged, but he had a light touch,' she told me. 'He was always trying to make the forest his home.'

Simard began following in her family's footsteps in the 1970s, heading off to the University of British Columbia to get a degree in forestry. She worked summer jobs in the bush and began noticing that forestry had changed dramatically since the days of her ancestors. Clear-cutting was common. The vast empty spaces left behind were replanted much like the megafarms that were replacing small family farms across North America, where crop diversity had given way to hundreds of acres of only one crop. Where there had once been great diversity in the native forest — a mixture of conifers like Douglas fir, lodgepole pine, hemlock, and spruce, plus broadleaf trees like birch, aspen, cottonwood, alder, and willow — the forestry companies were replanting a single, highly marketable variety of conifer in rows, much like a giant cornfield. As the conifer seedlings started to grow, some of the broadleaf varieties would try to stage a comeback, only to be sprayed to death with herbicides. The assumption behind this approach in both forestry and agriculture is that the crop — whether conifers or corn — will grow better without competition for water, nutrients, and sunshine from non-crop plants.

After Simard graduated and began working for a private

timber company in the 1980s, her initial revulsion at the ugliness of these plucked forests deepened to concern about their long-term health. 'The great thing about that job was that I was in the forest all the time and really got to know it,' she says of her work, which included laying out roads, planning the cuts and surveying afterward, planning the replanting, and even firefighting. 'But I was horrified by what was going on. There had always been outbreaks of the mountain pine beetle, but they became massive. Industry responded by clear-cutting whole valleys.'

In the end, she concluded, 'I had a good understanding that all was not well in the woods.'

Even though she still assumed that the prevailing wisdom was correct — that it made economic and maybe even ecological sense to replant the harvested forests in monocrops of pine, spruce, or fir — she started doing her own research to try to improve the health of the plantations. She had started grad school by the late 1980s and, still working for the timber company, set up hundreds of research plots in which she tested that wisdom. In plantations where native plants were starting to come back, she weeded them out in various combinations and degrees, looking for the point at which the conifers grew best as a result of reduced competition.

But that's not what happened.

'It was amazing,' she says. 'I was taking out alder and willow and birches and all these beautiful plants, and the conifers started dying. They kept dying. There was more disease and insect infestations. I kept looking at these plants and thinking something wasn't right.'

She went on to get a job as a government forestry scientist, tasked with figuring out how to make the plantations more successful. She told her new employer that based on her research, the plantations should be planted with a mixture of trees — and not just a mixture of conifers, but also with a sprinkling of the native broadleaves. 'That was anathema, the opposite of what they wanted to do,' she says. 'The government wanted industrial plantations with single-species rows like carrots and to produce two-by-fours, on and on into perpetuity.'

But her direct supervisor was interested in her research and urged her to get a doctorate so she could learn more about forest interactions. Still a government scientist, she enrolled at Oregon State University to study with ecologist Dave Perry, who had also come up with evidence that the presence of native broadleaves helped the conifers grow better. He was looking into the possibility that the trees were helping each other underground, perhaps via the long, silky strands of fungi. When we see mushrooms in the forest or elsewhere — I see hundreds in my Portland, Oregon, neighbourhood, especially after it rains — most of us don't realise that these are just the fruits of an organism that weaves a vast underground tapestry of fine threads called hyphae. There is an almost-inconceivable density of these hyphae: every time we take a step in the forest, Simard says, there are around 300 gossamer miles of them under each footprint.

At around the time Simard began her PhD work, British scientists had found that grassland plants in pots could be connected underground by mycorrhizae — meaning 'fungus roots', a composite structure formed by the fungi and the roots. Some 90 per cent of land plants are colonised by these mycorrhizae. In forests and other areas, conifers and some other trees associate with ectomycorrhizal fungi, which cover their root tips. Other trees and nonwoody plants associate with arbuscular mycorrhizal fungi, which pierce their roots. Although scientists originally thought that fungi were stealing the sugary carbon fuel that plants make during photosynthesis, they've known since the 1880s that in many cases plants are trading their carbon fuel to the fungi — which cannot make carbon through photosynthesis — in exchange for water and nutrients. Beneficial partnerships like this between different species are called mutualisms, and they occur in all ecosystems and probably involve every species on Earth. And they are hugely important, influencing everything from nutrient cycles throughout the biosphere to individual cells.

Simard wondered if the reason the conifers in her research plots failed when she removed the broadleaves was that they were not only connected through the soil, but also provided some benefit to one another through their fungal support systems. As part of her doctoral research, she tested this idea, injecting a radioactive form of carbon into young Douglas firs and birches — two of the 'early succession' trees that together colonise a bare or disturbed landscape — growing next to each other in a forest clear-cut. She mimicked fluctuations in sunlight and found that the birch shared carbon fuel through its ectomycorrhizal connection to the fir based on the amount of shade the birch cast on the fir: the more the fir was shaded, the more carbon the birch transferred to it. On the other hand, the western red cedar she had planted nearby and which connects to other plants via arbuscular mycorrhizae instead of ectomycorrhizae only received a tiny portion of the birch's largesse.

Later, one of her graduate students repeated this experiment over different seasons. As in the first experiment, when the birches leafed out in midsummer and the firs were shaded so much that they had a hard time photosynthesising, some of the birch's carbon flowed to the fir. Early in the spring and later in the fall, when the birch had no leaves and could not photosynthesise, some of the growing fir's carbon flowed to the birch. It isn't clear which organism — tree or fungus — orchestrates the transfer, but it is obvious that they cooperate to maintain a diverse, healthy forest community. As in the original experiment, western red cedars planted nearby received very little carbon from either the birch or the fir, because cedars are not part of the ectomycorrhizal network joining those other trees.

Simard's original experiment wound up being published in the prestigious scientific journal *Nature* in 1997 and set off an excited international clamour about this evidence of the 'wood-wide web'. Implications of Simard's work even reached Hollywood and inspired some of the thinking that led to director James Cameron's film *Avatar*, in which the roots of the revered Tree of Souls offer simultaneous connection to members of the Na'vi clan. Shortly after the publication in *Nature*, a reporter from the *Vancouver Sun* called the now-somewhat-famous Simard to ask her opinion of the Canadian Ministry of Forestry's practice of spraying broadleaf trees with herbicide in an effort to help the plantation conifers grow. Simard was still working for the ministry, finishing her PhD, and about to have a baby. She wearily replied, 'For all the good it does, they might as well be painting rocks.' That comment wound up on the paper's front page, and she almost lost her job. She decided government was the wrong place for her and moved on to academia, where she's been ever since.

Simard and an enthusiastic band of students — as well as a handful of other scientists around the world — continue to focus on the connections among trees and the underlying mycorrhizal networks and the implications these connections have for forests, especially given the hotter, drier climate to come. Scientists now think fungi have been building these water- and nutrient-distributing networks among themselves for a billion years and began extending them to plants a half-billion years ago, allowing plants to move onto the land from the waters and thrive there. Plants can't move around to search for food or water, but an individual fungus can snake its tiny hyphae far and wide, poking into the gaps between soil particles for water and siphoning it into the network. The hyphae also wrap around those soil particles to enzymatically snap off minerals and make them quickly available to both themselves and the plants in their networks — without that enzymatic action, those nutrients would only become available to plants through a slow soil-weathering process. When the seeds of trees drop into the forest duff and start to sprout, they indicate their eagerness to join in this mycorrhizal feast by exuding a chemical into the soil through their roots, inviting the fungi to connect.

What are the advantages for these young trees? Simard and her students began studying this by planting seedlings in a forest near a Mother Tree. Some of the seedlings had fine-mesh bags around their roots that didn't let the hyphae get through — so, they were prevented from joining the mycorrhizal network — and others had no mesh bags or other impediments to connection. They found that the linked-in seedlings were healthier, so much so that they survived at a higher rate than the solitary seedlings. They flourished even in the shadow of the Mother Tree, where it was hard to catch enough light to photosynthesise — and the lab determined that this was because the Mother Tree was pulsing extra carbon toward the seedlings.

Simard and other researchers around the world keep trying to tease out the value of these fungal connections: what are the benefits flowing through both ectomycorrhizal underground networks (root-like structures comprising fungus plus conifers and some other trees) and arbuscular mycorrhizal networks (comprising fungus plus most other plants)? Some of their studies are conducted in greenhouses with plants growing together in pots so the scientists can observe interactions away from the massive complexity of the real forest and eavesdrop on at least some of the dialogue among plants. Not a spoken language, but 'a language that is chemical,' explained Julia Maddison, a former student in Simard's lab. 'We don't know how a language that is not based on sound or sight works, but we know it's almost certainly more direct. When the chemical message arrives, it does something, without the need for a brain or other organ to interpret.' When they want to compare plant communities not connected to a mycorrhizal network with communities that are connected, researchers grow one group in pots with each individual plant wearing a fine-mesh sheath around its roots to prevent connection; in the other group, plants within the pots are allowed to connect normally. To prevent plants from communicating with each other via airborne volatile chemicals — and plants do this regularly — they cover the tops with bags.

These accumulated studies show that the mycorrhizal networks offer trees and some other plants more than just carbon, water, and mineral nutrients. They also function as a forest early-warning system: when attacked by insect or fungal pests, trees emit a sort of chemical scream into the underground networks, prompting other plants to produce chemicals that make them less appetising targets for the pests. Other benefits are still shrouded in mystery. A study by Swiss scientist Florian Walder of flax and sorghum grown together showed that the flax shared very little of its carbon with the network and withdrew huge amounts of nitrogen and phosphorus, whereas the sorghum dumped in a lot of carbon but didn't seem to be taking much of anything. Still, both plants grew better together than either did when growing only with its own kind, so they were clearly getting something from each other via the network — the science just couldn't pinpoint what it was.

'We just haven't looked hard enough,' Simard told me. 'All these interactions are fluid, changing all the time, happening all at once. If we could look at the whole suite of interactions, we'd find out there is something else they're getting out of the relationship.'

As tree seedlings grow, they keep adding mycorrhizal connections to the forest at large. They might associate with only twenty species of fungi when they're young, but as they mature into large trees, they can build underground partnerships with hundreds of fungal species. Simard's student Kevin Beiler came up with an ingenious method of tracing the fungal connections among Douglas firs in a thirty-by-thirty-yard plot containing sixty-seven trees in an old-growth forest near Kamloops, Canada. Beiler took DNA samples from the Douglas firs, then drilled into the soil to track the distribution on tree roots of two sister species of fungi from the genus *Rhizopogon*. The study revealed mycorrhizal networks that linked trees old and young, although the younger trees had far fewer connections than the older trees. One standout elder was connected to forty-seven other trees through its mycorrhizal connection with eleven individual fungi from the two species.

That elder is presumably connected to more trees of other species and to another 100–200 species of fungi that weren't being tracked by the study, all operating their own pipeline of goods, services, and news — in the form of chemical messages — from the forest at large.

To make an analogy to humans, this tree in Beiler's study is like my friend Lori here in Portland, or my friends Linda and Karen back in Cleveland, where I used to live. None of them can go anywhere in their respective cities without seeing someone they know — someone they taught, or worked with, or lived near, or dated the brother of. Through those women, you can probably find anything or anyone in those cities. You connect with its history and life and current moment. But the comparison is meagre, because the trees and fungi are engaged in an ongoing dialogue that also sustains life. We can't do that, despite our many tech-facilitated links.

Just as there are many tree species in a forest, there are also many species of fungi — at least fifty different kinds per acre, Simard says, even in a modest urban forest like the ones not far from my house in Portland. There are thousands of fungal species that can associate with Douglas fir, and they likely all offer a slightly different partnership to the tree. Just as trees can be early- or late-succession participants in an ecosystem — some find their niche early, like Douglas firs and birches; others only find their niche later, after those early colonisers have begun turning a bare, burned, or degraded landscape into a functioning biological community — there are also early- or late-succession fungi, which tend to associate primarily with the oldest trees in the landscape. The *Rhizopogon* of Beiler's study is a late-succession fungi genus associated with the bigger, older Douglas firs that Simard is focusing on for the Mother Tree Project. While Simard and her lab staff don't know everything this fungus does, they know it's especially great at taking up water. Thus, the Mother Tree connected to those forty-seven other trees from Beiler's study can keep offering them sips of water in a drought. And of course, that tree is likely connected to individuals from another 200 fungal species and is also connected to other tree species and their mycorrhizal networks. They are all whispering to each other in chemical sentences, giving to the forest community and taking from it in ways scientists may never be able to figure out.

The overlapping layers of connection are staggering and possibly infinite. 'Everything is connected to everything else,' Simard says. 'We find nested networks that could cover the landscape. When we lay down a pipeline or make a road or some other cut into the earth, then we sever the networks. But minus those disruptions, it's hard for me to imagine that it's not all linked.'

Simard and her lab have been primarily studying the interactions among trees and fungi, but the web of forest relationships is much bigger than that, of course. Animals are part of every landscape, part of the symphony of interactions that create an ecosystem, but too much of our science is siloed into looking at all these living things in isolation from each other. While I was writing this book, Allen Larocque, one of Simard's students, was investigating the impact of the Pacific salmon on these northwest forests.

Every year, millions of sexually mature salmon leave salt water for fresh water, returning to the rivers and streams where they hatched and spent the first part of their lives. This mad dash to spawn takes the salmon hundreds and sometimes thousands of miles from the ocean, even into mountain streams high above sea level. Most don't make it: some starve on the way there, some sicken in polluted waters near cities, some are caught by fishermen, and many are seized and devoured by bears, otters, and eagles.

We're used to thinking of salmon as a gift to eager inland creatures, but few of us realise that the salmon runs are also a huge nutrient pulse to forests and other landscapes. The primary nutrient they confer upon the land is nitrogen, which is essential to every living thing. Animals need it to make protein, DNA, and other important compounds; plants use it for all that plus it's a major component of chlorophyll; we're all hungry for it and dwindle without it. Nitrogen is the major gas in our atmosphere, but neither animals nor plants can use it in that form — it has to be 'fixed' in another chemical form, which happens naturally via bacteria or lightning strikes. Forests in the Pacific Northwest accumulate nitrogen in a variety of ways. Some are dotted with plants like the red alder tree that curry relationships with certain kinds of bacteria, forming special nodules in their roots where these bacteria fix nitrogen for themselves and the plant — an energetically demanding task that creates a form of nitrogen that lasts 400 years — and the trees provide them the energy to do this in the form of carbon fuel. Sometimes nitrogen drops to the forest floor from the branches of trees, where lichen host bacteria that fix nitrogen. But often the forest's biggest single source of nitrogen is from the dead bodies of salmon. Scientists call their contribution the salmon shadow.

'Kind of sounds like something from Mordor, this shadow spreading over the land,' laughs Larocque. 'It affects the whole landscape, even kilometres from the river. But the biggest effect seems to be 100 to 200 metres away.' That deeper salmon shadow changes the forest dynamics and influences what kinds of vegetation, insect, and bird communities thrive there.

Larocque had been working on this project for three years when I talked to him and had taken several trips to streams where the salmon were leaping. Where there are salmon, there are also bears. During one trip, Larocque and his colleagues rounded a bend in the river to see a salmon lying on a rock, newly gashed open by powerful claws, its heart exposed and beating. They stepped away for a few minutes and while their heads were turned, the bear snatched away its prey. 'The whole forest smells like a slaughterhouse when the salmon are running,' Larocque says. 'There are carcasses everywhere.'

And those carcasses are loaded with nitrogen. All plant and animal bodies contain nitrogen, which clusters in the proteins — meaning that protein-rich food is also nitrogen rich. In plants, proteins are concentrated in the most biologically active parts, like the root ends and the fast-growing shoots. 'This is why herbivores prioritise fresh green foliage over the older foliage,' Larocque explained. 'The fresh green tips have higher nitrogen content, as well as being tastier, I imagine.' Nitrogen thus accumulates in the food chain, and the animals that eat other animals — which are protein rich — have more nitrogen in their bodies than strict plant eaters. Even humans who are carnivores have more of a certain form of nitrogen — an isotope called 15N, which is heavier than ordinary nitrogen — than do humans who are vegan. Salmon are carnivores that accumulate a lot of 15N over their lifetimes. Thus, the salmon shadow on the land has a specific signature — that of 15N — making it easy for scientists to measure.

Some of this salmon-derived nitrogen makes its way deeper into the forest via bears and wolves that drag their prey to their favourite picnic sites, eat parts of the fish, and discard the rest. Teresa Ryan — who's also working on the 'salmon forest' project — says that these carnivores can carry as many as 150 fish per day into the woods during the salmon run. But the Simard lab suspects there is a larger nitrogen-spreading mechanism at work — namely, the forest's many overlapping mycorrhizal connections.

Larocque has begun to test this by first measuring the natural abundance of 15N in the soil, fungi, and vegetation at his test site — it's there as the result of aeons of salmon bringing nitrogen upstream. He first deposits a fresh load of

15N onto the ground, simulating a rotting salmon. Then he takes samples of soil, fungi, and vegetation over time to see if he can catch this wave of heavy nitrogen moving through the forest's many webs of relationships.

Miles from the forests of British Columbia, a University of Washington scientist stumbled upon another relationship that provides nitrogen to plants, including trees in a forest. 'This was not a eureka moment,' says ecologist Sharon Doty. 'It was a "what the hell?" moment.'

Back in 2001, Doty was studying the stems of poplar trees as part of a project to conduct genetic engineering that would help the trees degrade environmental pollutants. But every time she'd prepare stem tissue for the first step of the process, what she describes as 'slimy bacteria' leaked out. She sequenced the DNA of the bacteria because she figured that if she knew exactly what it was, then it would be easier to kill. She was astonished when the bacteria turned out to be rhizobia — the best known of the bacteria that enter the root nodules of leguminous plants like clover and peas and produce nitrogen in exchange for carbon fuel. Until Doty's discovery, scientists overwhelmingly believed that these particular bacteria only performed this nitrogen-fixing service in such nodules. The editors of scientific journals were hard to convince otherwise; it took her more than three years to publish a paper on her discovery.

The rhizobia and other nitrogen-fixing bacteria were clustered thickly inside the poplars' branches, leaves, and roots.

'Trees can defend themselves against pathogens, but they're letting these microbes colonise them at a high density,' Doty says. 'We sometimes find millions of microbial cells per gram of tissue.' The rhizobia enter through cracks in the roots and migrate throughout the tree through its vascular system. Doty thinks the rhizobia may also land on a leaf surface via a breeze, or in drops of rain, or be conveyed by insects. Once there, the bacteria gain access to the inside of the leaves through the stomata, tiny holes in the leaves that suck in the carbon dioxide needed for photosynthesis and expel unneeded oxygen.

Doty and her colleagues went on to study whether the nitrogen-fixing microbes found in the poplar leaves would do the same job in other plants. Using a solution made from the poplar rhizobia, they soaked the seeds and sprayed the roots of a wide range of other plants — from crop plants like tomatoes, maize, and rice to forest plants like Douglas firs and western red cedars — and found that all got a growth boost from these nitrogen-fixing bacteria. 'I think this is a very ancient symbiosis,' Doty says. 'Conifers like Douglas fir diverged evolutionarily from flowering plants like tomatoes many millions of years ago. The fact that both of these plants allow the microbes to colonise and provide these profound growth impacts suggests that this is a way plants always got nutrients, not from primates like us providing them with chemical fertilisers.'

Some plants may need a boost from nitrogen-fixing microbes in their tissues more than others. Poplars tend to live in rocky areas with almost no soil, so they're unlikely to benefit as much from the kind of underground sharing that trees in the thick of the forest get. Poplars are also an early coloniser that can flourish in rocky, sandy landscapes that don't have a lot of other trees and, thus, not a lot of mycorrhizal pathways between them; it seems possible that they are able to do this because they're especially good at attracting microbial partners to provide their nutrients instead. Another researcher, Christopher Chanway from the University of British Columbia, has found an abundance of nitrogen-fixing microbes in the needles of lodgepole pine, another early coloniser. Doty finds it plausible that these early colonising trees may make it easy for other trees to get a start in the forest by sharing some of their nitrogen through the emerging mycorrhizal networks.

Doty also found that the bacteria living in the poplar tissues can do far more than just provide nitrogen to the plant. The bacteria cluster inside the leaves near the stomata, producing a hormone that closes them during a drought, thus helping the plant save water. Even with the stomata closed, the plant doesn't starve for carbon dioxide: as the microbes eat the plant's carbon fuel, they release carbon dioxide as a waste product, just as we do, and that keeps the plant fed until the drought crisis passes. When Doty soaked the seeds of rice and other plants in the rhizobia solution, they, too, became more drought tolerant. Other researchers have done similar work with salinity tolerance, which is a huge problem in agriculture as irrigation tends to make croplands become saltier, and found that the microbes made plants more salt tolerant.

'Microbes have this incredible power to help plants,' Doty says. 'It's a huge untapped resource.'

A huge untapped resource for us as we try to manage the environments we've disrupted and continue to disrupt for our food, water, wood, living spaces, and so on — but business as usual for the rest of nature, where living things engage in constant, complicated interactions without our notice. As I plunged into the Malcolm Knapp Research Forest with Simard and Roach back in September 2017, I couldn't help but believe that every quivering twig, every lip of fungus curling away from a fallen tree, every sprightly tuft of moss — everything — was throbbing with a dance of life carried on by multiple partners. Science is just barely beginning to catch on.

Not that we can fault the scientists. There is only the time and the money and the technology and the skills to address a sliver of what's going on in the natural world — more of our science is directed elsewhere, often toward the kind of science that leads to product development — and it's hard work. During my two days of participation, a lot of gear had to be schlepped into and out of the woods. Simard and Roach gave me a few things to carry, but Simard herself carried a bulging clown-car backpack stuffed with tools and fell more than once from the lopsided load. When we got to the site, they set their instruments down and looked dismayed. 'That's a lot of trees,' Simard said.

'A lot of trees,' Roach agreed.

I was confused. They're forest scientists — shouldn't they

be glad to see a lot of trees?

We got to work 'taming the site' — getting a baseline of what actually is there before woodcutters come in and conduct one of the four harvesting methods. First, Simard and Roach determined the exact centre of the circular site — it was right near a 1,000-year-old cedar that still towered fifty feet above us even though it was broken and burned — then began dividing the site into big pie-shaped wedges with white tape. We had to account for all the trees within the wedges — their general health, their girth and height, their species, and much more data — and it became my job to nail small, numbered metal tags into each one, just below where the tree would be cut. Thanks to the numbers — each of which corresponded to a slew of statistics Roach entered on the study paperwork, along with our initials as the source of the data — Simard, Roach, and future researchers would understand the variety and distribution of trees that had formerly covered the site. We would also estimate the amount of other vegetation at the site and carry away samples of soil and mosses.

As I started measuring and tagging the trees, wedge by wedge, I cottoned on to their dismay at the number of trees on the site. There *were* a lot of trees, mostly slender Douglas firs, hemlocks, and cedars that took root in the aftermath of the harvest nearly eighty years ago, plus a handful of bigger trees that had escaped it. I banged my fingers more than once as I tried to get the tags close to the ground, all facing in the same direction, and tried to keep the numbers sequential, although every once in a while, I'd realise I had missed a tree and would lope over to bang a tag into it. One of Simard's students, Katie McMahen, followed behind me, calling out data about each tree to Roach, who jotted it down. I was just thinking I had the hang of this modestly challenging task when I heard McMahen call out, 'I can't find number 543!' I went back and helped her look and together we found it — several yards from 542 and 544, and hard to locate in the maze of trees.

Simard was watching this from the side of a soil pit she was digging twenty feet away, laughing. 'You know, Kristin,' she called out, 'those tags are still going to be here in a hundred years and the researchers are going to be saying, "Who was this KO that pounded in the tags?"'

It was funny but also a little dazzling. Simard's old research plots from more than thirty years ago are still protected and studied. This, her biggest study of all, would be visited over and over again by scientists long after we were gone.

Later, I sat and watched Simard work on the soil pit, one of four for this site. She had laid a twenty-by-twenty-centimetre square on the ground and began removing the forest floor below. First, she pulled away the loose duff on top, then the so-called fermentation layer — where branches and needles and recognisable chunks of biology are decomposed by microorganisms — then the humus layer below, where those chunks are so decomposed that they're unrecognisable. She kept digging until she reached the hard layer of mineral soil at the bottom. In a pit dug earlier that day, the mineral layer was more than a yard below the surface. This one was about two feet down. She picked at some chalky material in the hard soil. 'Sometimes there are tiny shells, because all of this was once covered by seas,' she mused. Then she pulled a tiny yellow branch from the fermentation layer and dropped it in my hand. It looked like a piece of coral but bendable and fragile.

'Piloderma, one of the forest fungi,' she announced. 'It's one of the showier ones. There are probably a hundred species of fungi in this forest, but most are hard to see.'

Later, McMahen told me how she and Simard met. She'd been working at the Mount Polley mine in British Columbia for five years. During her tenure, the dam to a big tailings pond — containing the ground-up rock left over from mining — broke and spilled a 'slurry of toxic water and mud into Quesnel Lake, once renowned for being the cleanest deepwater lake in the world', according to a local paper. Simard was brought in to consult about degradation of the nearby forest, and the two met and found a common interest in landscape restoration.

McMahen's research isn't only applicable to catastrophic mine disasters; it could address the more ordinary disasters that cause degraded landscapes the world over. 'We're testing some methods that use ecosystem legacies and memories,' she told me. 'Those are pieces of the environment before it was disturbed. Many things survive — fungal spores, seeds — and all those little memories help the system recover to where it was.'

It's such a simple project. Part of it echoes the Mother Tree work: little trees will be planted near the edge of the forest in the midst of the spilled tailings, which she says are so degraded they're hardly even soil. Some of the trees will be planted behind a trench that will cut off their access to the mycorrhizal networks operating in the forest, and some will not be blocked. McMahen and others will see how much the networks can reach past the edge of the living forest into the dead zone to breathe vitality into the new trees. Other trees will get a handful of forest soil as they're planted. The idea is that some of the forest's memory might endure in that soil and give the seedlings a boost.

'Yeah, forest memory — that's anything biological with a blueprint in it,' Simard said when I asked her about the project. 'Anything that provides DNA to the next living community.'

Is there living memory and DNA for wholeness everywhere, even in our most degraded landscapes? Even in our most degraded relationships, including those with each other?

## Chapter 2

# We Need Better Metaphors

We run along a well-tramped path through stands of Rocky Mountain wildflowers that bloom and bobble over our heads, chasing the elusive Orange 78. She disappears into a cloud of white corydalis, the enormous country cousins of some dainty purple flowers in my garden back in Portland. Then Yellow 54 sails into view and we rush to follow her instead, angling through the surrounding green curtains of vegetation to keep up. She makes a beeline for the mountain bluebell.

Truly, it is a beeline: she is a small bumblebee that was trapped, tagged (with a yellow dot marked with the number 54 superglued to her back), registered, and DNA-sampled a few days ago for research purposes.

Biology undergraduate student Karen Wang drops the fine-mesh net she carries, whips out a small recorder, and leans in to observe Yellow 54 as she climbs into the long tubular blooms of the bluebell. 'One,' she says, referring to the bee's quick entrance into a blossom, 'and out. Two ... out. Three ... out. Oh, four was a secondary rob! Five was a secondary rob! Six was a secondary rob! Ah,' her voice drops. 'Now she's legitimate again.'

Bees and flowers have one of the most studied of all mutualisms. Flowers produce nectar to attract bees, which enter the flower and climb past the pollen-powdered anthers for a quick energy drink, sometimes mixing some of that nectar with pollen to feed their young. The bees emerge from the bloom dusted with the golden grains. Then, they move from one flower to another, in effect paying for that nectar with a sort of third-party mating as they move the pollen around and, in that manner, pollinate plants and enable them to make seeds.

But sometimes bees 'cheat' — a somewhat hilariously censorious way biologists describe the behaviour of organisms that fails to perform their role in the mutualism. That's what Wang spotted Yellow 54 doing among the wildflowers along the side of the road at the Rocky Mountain Biological Laboratory, where dozens of biologists spend their summers studying wildlife.

Yellow 54's job isn't as easy as it looks, North Carolina State University ecologist Rebecca Irwin explained to me later. Her lab is studying three of the nine native bumblebee species found in that part of the Rocky Mountains. She says that mountain bluebells and other commonly robbed flowers have narrow, tubular blooms, which can make it tough for a fat bumblebee to clamber all the way past the anthers to the nectar.

Bees sometimes cheat by chewing a hole through the base of the bloom to get the nectar, which can be a faster way to their sugary reward but one that often omits pollination. Some of the bees also commit what the researchers call a secondary rob, as Yellow 54 did. They cheat by entering through one of the holes gnawed by another bee. In either case, the plant has done a lot of work — pulled carbon dioxide out of the atmosphere through photosynthesis, converted it into a sugary carbon-rich liquid, and finessed some of that liquid into nectar — with no reciprocal reward, at least as far as Yellow 54 is concerned.

Most people think of bees as models of cooperative behaviour. However, researchers keep finding more and more evidence that they and other creatures, from fish to bacteria, sometimes cheat. A few mutualist partners are even more outrageous cheaters than the bees. Some butterflies in the *Lycaenidae* family have a mutualism with ants, in which the caterpillars secrete a sugary liquid through special organs to feed the ant, and in exchange, the ants protect the caterpillars from creatures that want to eat them. Occasionally, though, one of the caterpillars manages to mimic the smell of an ant larva or egg, and the ants will dutifully carry it back to their nest. There, they will feed the caterpillar mouth-to-mouth, as they do their own young, even as the caterpillar chows down on the other larvae in the nursery.

'I always think the ants must be so surprised when a butterfly finally emerges,' says biologist Megan Frederickson of the University of Toronto, who studies mutualisms between ants and plants in tropical forests. 'There are great videos showing butterflies having to hurriedly escape the ant nest because the ants all of a sudden realise there's an invader.'

Other living things have also evolved to snatch up the rewards of a mutualism with none of the costs. For instance, orchids: they are one of the world's largest plant families, but a third don't make nectar to attract insect pollinators. A few of these orchids trick male insects into penetrating their blooms by creating the odour or shape of a female of the same species of insect. The males thrash around trying to copulate, covering themselves with pollen. Then they spread that pollen to other orchids in an ongoing series of hopeless trysts.

I had been invited to the Rocky Mountain Biological Laboratory (RMBL, pronounced 'rumble') — located in Gothic, Colorado, in a 1,000-acre former mining town, it boasts 'the world's largest annual migration of field biologists' — by Judith Bronstein, a University of Arizona ecologist and editor of the 2015 book *Mutualism* (now the authoritative scientific work on the subject). She and her students head there in the summer to study beneficial interactions among organisms in this high-mountain valley, a great opportunity not only because RMBL's surroundings are gorgeous, but also because the visiting scientists can build upon the work of 9,000 other biologists who have been studying nature there since 1928.

Bronstein introduced me via email to several colleagues who would be studying mutualisms at RMBL. When I arrived, I had plenty of people to follow around the meadows and mountains as they set up test sites and returned to collect data.

The day with Wang and the other students chasing numbered bumblebees was fun as well as intriguing. If you're studying mutualisms, why focus on how they fail? Because, Bronstein told me later, looking at these failures gives researchers a better understanding of the overall costs and benefits of cooperation.

'Mutualism is very puzzling, because it's easy to see how partners could exploit each other,' Bronstein explained. 'It's costly for plants to make nectar, and they could make less of it and still get as much attention from bees — so why do they make so much nectar? And the bees we're studying can just chew a hole through the plant to get nectar — it's sometimes faster than entering through the floral opening. Why don't they all just cheat? Why do they cooperate as much as they do when it doesn't seem to be in their interest?'

Sometimes 'cheating' partners *are* punished for their unwitting naughtiness. Most bacteria living in the soil don't have the genetic chops to fix nitrogen in the nodules of plants like clover, peas, and alder trees, but they sometimes slip inside with the good-buddy mutualist bacteria and join in the carbon feast, without a nitrogen payoff for the plant. Flowering plants don't seem to punish bees that steal nectar from a hole in the side of a blossom, but plants are sometimes able to sanction freeloading bacteria. According to research from biologist Joel Sachs's lab at the University of California, Riverside, plants will kill some of their own cells if the non-fixing bacteria sneak inside them.

But most mutualist partners persist in offering the goods,

even when their counterparts fail to reciprocate. That appears to fly in the face of mainstream scientific thinking, which has cleaved to a belief in competition and selfishness ever since the time of Darwin.

Before my trip to the Rocky Mountains, I hadn't known that bees or other organisms slacked on their mutualism duties; in fact, I was barely acquainted with the term 'mutualism', even though I had written about the beneficial interactions among plants and soil microorganisms in my previous book. When I started the research for this book, I googled for other researchers besides Suzanne Simard who were studying cooperation in nature, and I discovered ecologist Douglas Boucher's 1985 book *The Biology of Mutualism*. I was struck right away by something he said in the first chapter: '(Mutualism) is an idea that has been reborn in the last decade. Never entirely absent from ecological thought, it none the less fell out of favour as modern ecology grew, and only since the early 1970s have we begun to find it important again. No one can be sure that its recent renaissance will not in turn fade away, but at least today it is an idea which is steadily gaining ground. Ecologists once again find it interesting.'

But when Bronstein's book came out thirty years after Boucher's, the study of mutualism was still lagging behind ecology's more-consuming interest in hostile interactions among species. While Bronstein found that a quarter of the articles about interactions among species published in major scientific journals from 1986 to 1990 were about mutualism —

and of course, that meant an overwhelming 75 per cent were not — she also noted that mutualism had 'paltry coverage' in the textbooks of the time.

Scientists and historians of science point to a number of reasons that beneficial interactions are understudied and underappreciated. Boucher, now retired from the Union of Concerned Scientists where he ran a program studying the impact of deforestation and agriculture on global warming, thought it might have something to do with the politics and social context of the twentieth century.

'During the Second World War and after, people lived in a social world in which competition was an overwhelming fact of life,' he told me back in 2015 (notably, before the nastiness of the 2016 United States presidential election and the seething social divisions that continue unabated). 'Now, with the fall of the Soviet Union and the end of the Cold War, there is more openness to the idea of cooperation in society and that reflects a willingness to see it in the natural world, as well.'

Bronstein's book notes that famed biologist and iconoclast Lynn Margulis — more on her later — argued that mutualism has been disregarded because men dominate the sciences and, since they are more aggressive themselves, are more likely to search for and find aggression in nature. Both Boucher and Margulis also noted that the idea of mutualism was championed early on by socialists and anarchists — more about this later, too — and thought that made Western scientists skittish. Another argument holds that competition is just more interesting to observe and study than cooperation — similarly, stolen UPS packages and flame-throwing tweets draw attention and inspire headlines, but the sight of neighbours lending tools to each other or people volunteering to help prisoners start gardens or clean plastic debris from ocean beaches don't. Many of us non-scientists assume that science focuses the beam of its intelligence on the issues and areas of inquiry that matter most, but, of course, science is like every other human endeavour: biases and cultural trends and big personalities, as well as the not-insignificant matter of funding, scatter the focus.

Beneficial interactions in nature weren't always disregarded by scientific thinkers, though. Boucher has an entire chapter on the history of thought regarding cooperation and competition in nature, and he points out that earlier Western civilisations' version of ecological theory was 'the balance of nature'. Ancient writers used examples of mutualism to show that nature was balanced so that there were never too many or too few members of a species. The ancient Greek historian Herodotus tells of a plover that plucks leeches out of the mouth of a crocodile, saying that 'the crocodile enjoys this and never, in consequence, hurts the bird'. Cicero and others offered similar stories, suggesting that humans could learn better behaviour from animals.

By the Renaissance, the balance of nature was seen as a hierarchy analogous to the structure of human society, Boucher explains. Plants and animals were thought to exist to feed humans and each other, and each organism had a preordained rule determined by God. The scientific revolution that began in the seventeenth century actually reinforced these ideas. The Swedish botanist Linnaeus, who developed the modern system of naming and classifying organisms, suggested that 'animals serve in the first place to preserve a due proportion among vegetables; secondly to adorn the theatre of nature and consume every thing superfluous and useless; thirdly to remove all impurities arising from animal and vegetable putridity; and lastly, to multiply and disseminate plants and serve them in many other respects ... Thus we see Nature resemble a well-regulated state ...'

The Industrial Revolution, which began around 1750, exploded the idea of that orderly and congenial natural world. Old social and political hierarchies and ways of life were upended in favour of new relationships among people as well as between people and the rest of nature. The fortunes of the new industrial barons soared while those of ordinary people crowding into the cities plummeted. One of the most prominent voices of the time was Thomas Malthus, a wealthy pastor and social economist who argued that human reproduction would always outstrip food and other resources, creating a struggle for existence that would cause most of the population to suffer — and that this struggle actually balanced society and was thus good for it. 'Malthus simply took one element of natural theology — the idea that deaths created harmony — and expressed it in terms of the industrial society developing around him,' writes Boucher. The nineteenth-century philosopher Herbert Spencer, who is now credited as the intellectual father of modern laissez-faire capitalism, also pointed to struggle and competition as a force that created progress. And although Karl Marx's sympathies did not lay with the wealthy, he also believed that struggle — that of workers against capitalists — would lead to a better society.

Charles Darwin was influenced mightily by these ideas as he developed his theory of evolution to explain the fossils of extinct species as well as the staggering variety of life forms seen everywhere in nature, arguing that all had descended from a common ancestor and that the ones with less successful traits had been weeded out by natural selection. He may have been predisposed by his station in life to see some groups or individuals as better suited for success than others, according to evolutionary biologist Kenneth Weiss, co-author of *The Mermaid's Tale: Four Billion Years of Cooperation in the Making of Living Things*. 'Darwin was a kindly person, but he was part of the gentry at the height of the British Empire and saw the Brits as better than anyone else,' Weiss says. 'He was a product of his time, as we all are.'

In Darwin's autobiography, he credits Malthus with inspiring a new theoretical direction: 'In October 1838, that is, fifteen months after I had begun my systematic inquiry, I happened to read for amusement Malthus *[...] on Population*, and being well prepared to appreciate the struggle for existence which everywhere goes on from long-continued observation of the habits of animals and plants, it at once struck me that under these circumstances favourable variations would tend to be preserved, and unfavourable ones to be destroyed. The results of this would be the formation of a new species. Here then I had at last got a theory by which to work.'

According to a delightful essay called 'What's the Point If We Can't Have Fun' by the late, great anthropologist David Graeber, Herbert Spencer 'was struck by how much the forces driving natural selection in *On the Origin of Species* gibed with his own laissez-faire economic theories. Competition over resources, rational calculation of advantage, and the gradual extinction of the weak were taken to be the prime directives of the universe.' After reading Darwin's *Origin*, Spencer developed the phrase 'survival of the fittest' to describe that competition. Darwin later borrowed the phrase for the fifth edition of *On the Origin of Species* in 1869 to explain how individuals with the best traits passed them on to the next generation, and eventually could change enough through the struggle to survive to form an entirely new species, while the evolutionary losers — without traits that ensured survival and reproduction — faded away.

These powerful ideas and metaphors about competition and progress-through-struggle were applied by Spencer and others to human society in the form of Social Darwinism, which affirmed the superiority of the wealthy and the deserved misery of the poor. But resistance to them was taking shape. Trade unions and associations through which Britain's working poor could pool their resources formed; in France, workers formed mutual aid associations, which, Boucher says, were a hotbed of socialist ideas. A working-class student named Pierre-Joseph Proudhon catapulted to the fore with his book *What Is Property?*, arguing for a kind of mutualism comprising workers' cooperatives that could ultimately, and nonviolently, replace capitalism.

Again, terminology would jump disciplines as Proudhon's word *mutuellisme* was applied to biology by the Belgian scientist Pierre-Joseph van Beneden. In an 1873 lecture, he said, 'There is mutual aid in many species, with services being repaid in good behaviour or in kind.' Both the idea and the word mutualism were taken up by biologists — and championed in Russia by Karl Kessler — but they got their real boost in 1902 when the Russian anarchist Peter Kropotkin wrote a best-selling book called *Mutual Aid: A Factor of Evolution*. Look him up online: that book and his many others, including *Memoirs of a Revolutionist* and *The Conquest of Bread*, are still in circulation. Yes, it was Kropotkin's incendiary stain on mutualism that may have spooked mainstream Western science.

Kropotkin has also been the subject of several books, including evolutionary biologist Lee Alan Dugatkin's *The Prince of Evolution*. 'Of all the people I've researched in the history of science, Kropotkin is the most interesting and one of the most important,' Dugatkin told me. 'He was the first person to formalise this idea of mutualism from an evolutionary perspective; he had the name recognition of a Richard Dawkins or Steven Pinker; he was chased by the

Russian police for decades, and he went around the world telling these stories about science and politics. This is someone who deserves to be known by a lot more people.' Truly, after reading Dugatkin's book and a few of Kropotkin's, I've decided that filmmakers looking for flesh-and-blood superheroes could have a cinematic contender in this man.

Kropotkin was born a prince in 1842 to one of the oldest and wealthiest noble families in Russia. 'Wealth was measured in those times by the number of "souls" which a landed proprietor owned,' Kropotkin wrote. 'So many "souls" meant so many male serfs: women did not count. My father, who owned nearly twelve hundred souls, in three different provinces, and who had, in addition to his peasants' holdings, large tracts of land which were cultivated by these peasants, was accounted a rich man.' Kropotkin could have gone on to similar bounty and a brilliant life within the court of the tsars. He had already charmed Tsar Nicholas at a ball when he was eight years old, and seven years later, he was chosen to join the tsar's Corps of Pages, an academy that prepared the children of nobility for the military and service to the court. But well before then, Kropotkin was beguiled by other interests: nature and radical politics.

Kropotkin's family spent summers in the countryside away from Moscow, and in his memoirs — written as a series of essays, in English, for the *Atlantic Monthly* and published by Houghton Mifflin in 1899 — he writes that his happiest childhood memories came from the journey to the summer estate, which took them five miles through a pine forest. 'The sand in that forest was as deep as in an African desert, and we went all the way on foot, while the horses, stopping every moment, slowly dragged the carriages in the sand,' he wrote. 'When I was in my teens, it was my delight to leave the family behind, and to walk the whole distance by myself. Immense red pines, centuries old, rose on every side, and not a sound reached the ear except the voices of the lofty trees. In a small ravine a fresh crystal spring murmured, and a passer-by had left in it, for the use of those who should come after him, a small funnel-shaped ladle, made of birch bark, with a split stick for a handle. Noiselessly a squirrel ran up a tree, and the underwood was as full of mysteries as were the trees. In that forest my first love of Nature and my first dim perception of its incessant life were born.'

At the same time that he was falling in love with nature, Kropotkin was being educated in radical ideas by one of his tutors, a Moscow University student named Smirnoff. Against a backdrop of war in the Crimea and dissatisfaction with the 'iron despot' tsar, Smirnoff enlisted his young student in hand-copying manuscripts of censored writers like Pushkin and Gogol and then distributed them. Smirnoff showed the young Kropotkin around Moscow, pointing out the homes of both political exiles and writers. Another tutor also encouraged Kropotkin's subversion, writes Dugatkin, and after he 'shared stories of noblemen renouncing their titles during the French Revolution, Peter ceased to refer to himself as Prince, henceforth signing his name as P. Kropotkin'. The world didn't forget, however: in the many newspaper articles and pronouncements that followed his books and speeches and many dramatic exploits, he was often referred to as 'Ex-Prince Kropotkin'.

When Kropotkin finally entered the Corps of Pages at the age of fifteen, he was ill-suited philosophically to prosper there. Despite some admired teachers and his own ferocious love of reading and study, he loathed the petty bullying carried on by the master of the school, 'a despot at the bottom of his heart, who was capable of hating — intensely hating — the boy who would not fall under his fascination.' Kropotkin was often disciplined for insolence and his refusal to follow what he considered to be stupid orders, and was once punished with ten days in a dark cell without his precious books. Nonetheless, he excelled so mightily in his studies that he was appointed sergeant of the pages, which granted him authority over the rest of the pages as well as an intimate connection with the tsar himself.

That greater intimacy, usually a stepping-stone to further honours and career advancement, actually wound up severing his loyalties to the imperial state, as he saw firsthand the tsar's cruelty and contempt for ordinary people. In his *Memoirs*, Kropotkin writes of attending the tsar — in his new position, Kropotkin had to keep pace with the tsar at everything from balls to military processions — as a peasant fell on his knees and offered the tsar a petition. 'I was close behind him [Tsar

Alexander II],' Kropotkin wrote, 'and only saw in him a shudder of fear at the sudden appearance of the peasant, after which he went on without deigning even to cast a glance on the human figure at his feet … I took it [the petition], although I knew that I should get a scolding for doing so. It was not my business to receive petitions, but I remembered what it must have cost the peasant before he could make his way to the capital, and then through the lines of police and soldiers who surrounded the procession. Like all peasants who hand petitions to the Tsar, he was going to be put under arrest, for no one knows how long.'

As his time in the academy came to a close in 1862, Kropotkin and the other pages were expected to pick a branch of the military for the next stage of their careers. He wanted to go to the university, but knew his father — a military man who had never fought but much enjoyed all the ceremonial preening — wouldn't allow it. He shocked everyone and dismayed his family by requesting a position in a new mounted Cossack infantry in Siberia — Russia had recently annexed the Amur region — where he hoped to carry on scientific observations similar to those of his hero, Alexander von Humboldt, as well as advance political reforms among the workers there.

Just nineteen years old, Kropotkin commenced what turned out to be five years of travel, totalling 50,000 miles, 'in carts, on board steamers, in boats, but chiefly on horseback' through the cold north. The trip would add momentum to the leftward drift of his political views, as one of his duties as a member of the tsar's army on his first expedition was to write a report on prisons in the Amur. He was predictably appalled by the harsh conditions and by bureaucratic ineptness. He met political exiles languishing there, and one — the poet and advocate for women's rights M.L. Mikhailov — gave him a copy of Proudhon's anarchist tract.

Another of the Siberian expeditions — this time, the assignment was to find a route between gold mines — provided a journey that would prove as foundational to Kropotkin's thinking about nature as Darwin's travels to South America on the HMS *Beagle* were to *On the Origin of Species*. The budding Russian intellectual had read Darwin's classic not long after it came out in 1859 and eagerly looked forward to studying, in the steppes and the mountains, the struggle for existence that was key to Darwin's thinking, but he found his expectations thwarted. Instead, Kropotkin was struck immediately by how valiantly living things had to struggle against the ferocity of nature and how they often clustered together to withstand it. He observed warfare and extermination among animals but was surprised to see 'there is, at the same time, as much, or perhaps even more, of mutual support, mutual aid, and mutual defence amidst animals belonging to the same species or, at least, to the same society. Sociability is as much a law of nature as mutual struggle. Of course it would be extremely difficult to estimate, however roughly, the relative numerical importance of both these series of facts. But if we resort to an indirect test, and ask Nature: "Who are the fittest: those who are continually at war with each other, or those who support one another?" we at once see that those animals which acquire habits of mutual aid are undoubtedly the fittest.'

The kind of cooperative behaviour *within* a species that Kropotkin describes here — and calls 'mutual aid' — is very different from the mutualisms between different species, like that between flowers and bees. As Judith Bronstein told me, 'Within-species cooperation and between-species cooperation have things in common, but they aren't identical ecologically, and they involve very different evolutionary processes.'

Kropotkin not only observed animals cooperating for protection and the procurement of food, but also marvelled at the ways in which they sought each other's company just for pleasure. 'Sociability — that is, the need of the animal of associating with its like — the love of society for society's sake, combined with the "joy of life", only now begins to receive due attention from the zoologists,' he wrote. 'We know at the present time that all animals, beginning with the ants, going on to the birds, and ending with the highest mammals, are fond of plays, wrestling, running after each other, trying to capture each other, teasing each other, and so on. And while many plays are, so to speak, a school for the proper behaviour of the young in mature life, there are others, which, apart from their utilitarian purposes, are, together with dancing and singing, mere manifestations of an excess of forces — "the joy of life", and a desire to communicate in some way or another with individuals of the same or of other species — in short, a manifestation of sociability proper, which is a distinctive feature of all the animal world.'

In 1867, Kropotkin left Siberia for St Petersburg, where he become a student at the university. His ostensible area of study was mathematics, but he immersed himself in politics. While in Siberia, his views had shifted even further to the left, not only because he met political exiles there and read Proudhon; he found that the mutual aid he so admired in the rest of nature also flourished among the humans there. In these scattered settlements far from government bureaucracy, he observed people treating each other with care and decency, unassisted and unsullied by the apparatus of the state. That stateless cooperation became his ideal. 'The fact that animals displayed mutual aid — and did so in the absence of anything remotely like government — suggested its deep biological roots,' writes Dugatkin. 'Kropotkin felt that the process of evolution had favoured mutual aid in animal populations, and if he had to put a political label on the way that these animals behaved, it would be "anarchy".'

If Kropotkin were among us today, he likely wouldn't be wearing the black bandanas of the antifa; although Kropotkin wanted the capitalist and imperial states to come tumbling down, he thought nonviolent strikes and shutdowns would do the trick. Nonetheless, his views and activities would soon get him into trouble. He became involved with a group of students who were distributing radical texts to the peasants and educating them about European trade unions. These students also went to the villages themselves, helping doctors and midwives, teaching — anything to be close to the peasants. The tsarist state watched these activities with alarm and was already making arrests. Kropotkin feared that he might be swept up, but one of his scientific interests kept him from seeking safety underground.

Before he went to university, he had undertaken one more major excursion for the tsar, exploring Manchuria through a border region that Russia and China were grappling over. On this trip, he became fascinated by the geology of mountains and made what he considered his first contribution to science: a study of the orientation and formation of the mountain ranges of Asia. This drew the attention of the Russian Geographical Society, which sent him on an expedition to study glaciers in Finland and Sweden. He had been invited by the society to deliver a paper on the origins of the ice age, and he did so. Hours later, he was arrested and thrown into prison. While in prison, he finished *Researches on the Glacial Period*.

Kropotkin's health began to suffer after two years in prison, and he was transferred to the prison hospital. As he improved and the authorities allowed him more outside contact, he and his friends devised a madcap plan — a caper worthy of Indiana Jones, employing red balloons and violins as secret signals — that busted him out of prison and spirited him away to England. So began his life as an exile. He got a job in London editing books for the new journal *Nature*, but he pined for the international anarchist movement, which was flaring more brightly on the Continent. He moved to Switzerland and got involved in both geographical work and anarchist politics there, but was thrown out of the country after the tsar's assassination, even though he had nothing to do with the group that carried it out. He and his wife settled in France, but he was surrounded by Russian spies and arrested by French authorities for being a member of an anarchist-socialist organisation, and was sentenced to five years in prison. A cluster of international luminaries petitioned for his release, including British members of Parliament, the novelist Victor Hugo, leaders of the British Museum, and editors of the *Encyclopedia Britannica*. The French government finally relented, and the freed Kropotkin returned to England, where a nascent anarchist movement was finally taking off. Although anarchists were largely feared and despised, Dugatkin writes that 'Kropotkin was seen as something of the exception to this rule, as many in Britain were fond of his ideas on mutual aid and enamoured with the audacity of his escape [from prison] … he continued to preach his ideas on everything from mutual aid to anarchy and socialism.' By this time, Kropotkin had renounced his family fortune and was making his living as a writer. He was offered a chair in geology at the University of Cambridge if he abstained from political activity, but he turned it down.

Still, he paid close attention to the conversations and controversies in the scientific community. In 1888, he reacted with fury to an essay by the scientist Thomas Henry Huxley, a prominent devotee of Darwin's who loathed anarchists and had refused to sign the petition for Kropotkin's release from the French prison.

Huxley wrote the article several years after Darwin's death, an event that likely still grieved both men. But Kropotkin had always been leery that Darwin's writings might be misconstrued by his followers and encourage an overly harsh view of life. He didn't believe Darwin actually intended that view. In fact, he writes, Darwin intimated that 'the fittest are not the physically strongest, nor the cunningest, but those who learn to combine so as mutually to support each other, strong and weak alike, for the welfare of the community.' He quotes Darwin from *The Descent of Man*, written twelve years after *On the Origin of Species*: 'Those communities which included the greatest number of the most sympathetic members would flourish best and rear the greatest number of offspring.' Kropotkin sums up what he sees as a transition in Darwin's views: 'The term [struggle for existence], which originated from the original Malthusian conception of competition between each and all, thus lost its narrowness in the mind of one who knew Nature.'

But Huxley's essay took Darwin's views down that harshest, most uncongenial path. 'From the point of view of the moralist,' Huxley wrote, 'the animal world is on about the same level as a gladiators' show. The creatures are fairly well treated and set to fight: whereby the strongest, the swiftest, and the cunningest live to fight another day. The spectator has no need to turn thumbs down, as no quarter is given.' In the same article, Huxley linked this gladiator show to early humans: 'the weakest and the stupidest went to the wall, while the toughest and shrewdest, those who were best witted to cope with their circumstances, but not the best in another way, survived. Life was a continuous free fight, and beyond the limited and temporary relations of the family, the Hobbesian war of each against all was the normal state of existence.'

Kropotkin asked the editor of the magazine that published Huxley's essay if he could write a rebuttal, and the full-length essay he submitted became the first instalment of a series that he would later turn into his best-known book, *Mutual Aid*. He continued refining his counterargument over the years: that struggle and competition certainly existed in nature, but that mutual aid and sociability were equally strong and may have the greater role in shaping evolution. There was great demand to hear him speak about these and other ideas. He made two major trips to North America in 1897 and 1901, moving across the continent to accept one invitation after another, speaking on a variety of subjects — about geology and glaciers in Toronto; about Siberia in Washington; about mutual aid and trade unionism in New York City; and three different lectures about the socialist movement in Europe, Christianity, and morality in Boston, returning to that city later to give a series of lectures on Russian literature. One speech on anarchism drew a crowd of 4,000 in New York. Kropotkin was planning a third lecture tour in North America, but public sentiment shifted dramatically after President William McKinley was assassinated by an anarchist in Buffalo. The mood in the United States turned against anarchists, and rumours circulated that Kropotkin and his friend, anarchist Emma Goldman, had something to do with the murder. Congress then passed the Immigration Act of 1903, adding anarchists to its list of banned newcomers.

Kropotkin continued writing about mutual aid, evolution, Darwin, anarchism, and more from England, but when Russia's Romanov dynasty was toppled in 1917 by the kind of strikes and massive protests he had long championed, he returned to Russia. He was greeted by thousands of people and offered a position in the provisional government, but he declined — it was a far better government than the tsars', but it was still a state. Later that year, Lenin and the Bolsheviks took over, and the increasingly centralised state became even more antithetical to Kropotkin's beliefs. He and his wife moved to a small house well outside Moscow, where he finished up his last book — *Ethics: Origin and Development*, on the role of mutual aid among humans and other animals — and assumed a modest role with a peasant-run cooperative, curating their local geology collection. When he died in 1921, his body was sent to Moscow. His family declined Lenin's offer of a state funeral; instead, anarchist groups funded the funeral, which was attended by thousands. Among the mourners were representatives of organisations — trade unions, anarchist groups, scientific associations, and literary societies — their colourful banners testifying to the breadth of Kropotkin's passion and influence.

Really, what a guy!

Yet, he's a guy most of us have never heard of. Anarchists still read and revere him; intellectuals like David Graeber rope him into their musings once in a while; but he has been largely missing from the literature of science. Although that's where Dugatkin found him while he was researching the evolution of cooperation among animals in the 1980s. He saw citations to Kropotkin that initially confused him, not realising that the famous anarchist had also been a leading scientist.

Dugatkin became obsessed with Kropotkin, reading all his work and just about everything that other people had written about him, visiting the large Kropotkin archives in Amsterdam, attending a conference in Moscow about him, and, with a colleague, translating his diary into English. He also studied the University of Chicago biologists who built upon Kropotkin's ideas in the 1930s and '40s, as well as the emerging field of sociobiology in the 1960s, which largely agreed that natural selection could favour mutualism and cooperation, but he concluded that the ways Kropotkin thought that happened didn't stand up to a detailed theoretical analysis. 'People erased the discussion of Kropotkin and his work altogether because of that second point, and his work was put on the back shelf,' Dugatkin says. 'But since 2000, there is a recognition that we tossed him away in the whole when we shouldn't have. The notion that natural selection could favour cooperation, which was a radical notion during his time, has stood the test of time.'

But in the meantime, other biologists piled on new

metaphors. One of the most powerful was 'the selfish gene', which was the title of evolutionary biologist Richard Dawkins's hugely influential book from 1976, considered by some to be the first scientific bestseller. Dawkins argues that the process of evolution is driven by genes, and that everything living organisms are and do merely serves to create the best vehicle for carrying those genes into the future. 'We are survival machines,' he argues in the book, 'robot vehicles blindly programmed to preserve the selfish molecules known as genes.' In part, Dawkins and the neo-Darwinist scientists whose work he drew from were targeting what Darwin worried was a weakness in his theory — the existence of altruism between members of the same species. A classic example: in beehives, female worker bees forego reproduction in order to care for the queen bee's offspring, thereby sacrificing their own apparent self-interest for that of the hive. Dawkins's book argued that there was no theoretical weakness in such cases if one regarded genes as the driver of evolution: in his analysis, altruism evolved because it helped bump the genes an individual shared with kin into the next generation. Of course, Dawkins intended his title as a metaphor — he didn't mean to suggest that genes have the consciousness that selfishness implies — but his metaphor combined with the ones inherited from Malthus, Spencer, and Darwin to paint a picture of a living world driven by selfishness and struggle.

The title became more and more controversial over the years, and Dawkins has often said that he might have named the book otherwise — The Immortal Gene, maybe. His defenders have batted away complaints by repeating that the title was metaphorical. But, of course, metaphors matter, especially when they come from leading scientists trying to describe reality and become a widely repeated meme among the greater population. Metaphors shape our views of the world and help us create guidelines for how we behave and relate to others, for how we interpret the world around us and for what we expect in the future. 'Different cultures construct core metaphors to make meaning out of their world and these metaphors forge the values that ultimately drive people's actions,' explains Jeremy Lent in his book *The Patterning Instinct: A Cultural History of Humanity's Search for Meaning*.

Dawkins had come up with an especially bleak twist on Darwinism, according to anthropologist David Graeber. Darwin had theorised that even tiny differences among members of a species could determine fitness. Given this belief that every difference matters, the scientists who followed in Darwin's footsteps have assumed that selection focuses with laser-like precision on those differences. 'The neo-Darwinists were practically driven to their conclusions by their initial assumption: that science demands a rational explanation, that this means attributing rational motives to all behaviour, and that a truly rational motivation can only be one that, if observed in humans, would normally be described as selfishness or greed,' Graeber writes. 'The neo-Darwinists assumed not just a struggle for survival, but a universe of rational calculation driven by an apparently irrational imperative to unlimited growth.'

Such metaphors of constant struggle and selfishness are like a dark lens slipped over the eye, changing the colours of the world around us to grey. Or like the glass splinter in the original Snow Queen story by Hans Christian Andersen, a far different tale than the wildly popular adaptation, *Frozen*, that Disney would produce more than a century later. In 'The Snow Queen', a devil creates a magic mirror that distorts everything reflected in it, showing only the world's flaws and ugliness, and obscuring its goodness. When the mirror breaks, a glass splinter lodges in the eye of a boy who suddenly finds the world around him bleak and wanders off, forcing his spurned friend to begin a heroic journey to rescue him from a land of ice. Similarly, metaphors of constant struggle and greed convince us that justice is likely to be thwarted, that benevolence is suspect, and that we — as the supposed apex organisms in this long churn of evolutionary history — are only fulfilling our biological destiny as we eat up the rest of the planet. They suggest that life is all a zero-sum game, in which a benefit for one is a loss for another. How can we not feel utter hopelessness at improving our culture and repairing our relationship with the rest of nature if we're locked into these metaphors?

Even when people committed to these metaphors of struggle and selfishness are presented with an example of extreme generosity — say, when a person jumps into a pool to save a person who's drowning — they often find a way to explain it that emphasises struggle and selfishness. 'The hyper-Darwinists just say cooperation is competition in disguise,' says evolutionary biologist Ken Weiss, who actually did dive into a pool once to save a drowning woman. 'They'd say I did that because at some future time, she or someone else might save me. The technical term for that is bullshit. I saved her because I have empathy.'

Are these metaphors and mindset not only unpleasant, but wrong scientifically? Weiss and his co-author — his wife, anthropologist Anne Buchanan — think that's the case. Their book argues that a huge number of components interact to create the diversity of species in the world. Of all these components, Weiss says, 'cooperation is a fundamental principle of life, arguably much more pervasive and important than competition because it happens at all levels, all the time, from minuscule intracellular spaces to grander ecosystems, instantaneously as well as over evolutionary time. Your body is made of billions and billions of cells that have to interact for a common cause, and I would call that cooperation. Without that, you wouldn't be a multicellular organism.' Certainly, some members of species fail to survive because they are ill-suited to withstand the demands of their environment, but Weiss and Buchanan call this the failure of the frail — meaning that an organism doesn't have to be the fittest to survive and reproduce. In fact, organisms in the same species can have a range of variation and all still be fit enough to survive and reproduce.

And despite the neo-Darwinian reflex of looking at ourselves and the species around us today as winners in the long-distance race called evolution, fitness doesn't always ensure that a species will survive. Chance rolls the dice constantly, and luck, Weiss argues, can have a greater impact on survival than fitness. Volcanic eruptions, hurricanes, wildfires, tsunamis, floods, and any number of disasters or external changes carve new grooves in the populations of living things, and the organisms that successfully move their genes into the next generation aren't necessarily the fittest organisms; they're the luckiest. Or, as Weiss quips, it's the 'survival of the safest'.

The view of evolution as a process driven by competition and selfishness was thoroughly shaken up in the 1960s by evolutionary biologist Lynn Margulis. She was fascinated by the microbial world; her daughter Jennifer Margulis says that she called herself a 'spokesperson for the microcosm'. In a paper that was rejected by fifteen scientific journals before it was published, Margulis pointed back to the earliest days of life on Earth and argued that single-celled organisms (bacteria appeared some 3.8 billion years ago) made a dramatic leap into multicellular complexity via symbiosis some one and a half billion years ago. In this hypothesis, two different microorganisms lived in community — as bacteria still do, forming slimes and mats in which millions live and interact with each other — and merged to form a new and more internally complex kind of cell called a eukaryote. Those eukaryotic cells themselves went on to form symbioses to become multicellular organisms. Eukaryotic cells contain bundled structures outside the cell nucleus — mitochondria, which produce energy, and, in plants, chloroplasts, which drive photosynthesis — and Margulis argued that both are remnants of formerly free-living bacteria. This line of thinking stretched all the way back to the work of early twentieth-century Russian scientists who embraced both Kropotkin and Darwin but were ignored in the West at the time. When Margulis presented her work, it took ten years for most of the rest of science to stop damning her as a heretic, but her arguments are now widely accepted. Every bit of our bodies, as well as those of other animals, plants, and fungi, are made of eukaryotic cells — these minuscule bundles of cooperation that transformed Earth more than anything except for the emergence of life itself.

A friend who heard I was working on this book sent me a marvellous essay by neuroscientist Kelly Clancy for *Nautilus* magazine called 'Survival of the Friendliest', which introduced me to the concept of 'relaxed selection'. As Clancy points out in the essay, natural selection — the weeding out of members of a species that have traits that make them less likely to survive, and the resulting surge of others with more helpful traits — can be 'relaxed' by events outside an organism's control, like a drop in the number of predators or the sudden increase in a food source or a long spate of fine weather, and this relaxation allows organisms the freedom to change and grow in new ways. But selection can also be relaxed by the actions of the organisms themselves. 'Evolution isn't just selecting for bodies,' Clancy explained to me. 'It's selecting for behaviours, postures, mating dances, habitats. It's beavers making dams, and humans making cities. It operates on a cultural level.'

The ancient fusion of microorganisms that created eukaryotic cells was surely an example of relaxed selection: the two single-celled organisms struck a bargain in which one found a safe environment inside the other, and the host acquired an onsite energy source. 'Here, evolution is not a weapons race, but a peace treaty among interdependent nations,' Clancy writes. The new eukaryotic cell was given the freedom by this union to expand its numbers and be more biologically creative — a creativity that ultimately led to greater complexities like us and all the animals, plants, and fungi in our world.

Relaxed selection goes on all around us, offering similar freedoms to other organisms. Clancy's essay points to one fascinating example that goes on in oceans around the world. There, the cyanobacteria *Synechococcus* and *Prochlorococcus* live side by side in floating communities. Both feed themselves via photosynthesis, turning sunlight and carbon dioxide into a sugary fuel. But as they photosynthesise, they create a toxic byproduct and need to protect themselves by squirting an enzyme into the water to counter it. It takes a lot of energy to produce the enzyme, and only *Synechococcus* has the gene to do it. But *Prochlorococcus* benefits all the same, as the enzyme *Synechococcus* produces wafts through their common soup and becomes a shared good. *Prochlorococcus* doesn't need to waste any energy producing the enzyme and, instead, can focus more energy on reproduction. And it does a champion job of it, benefitting the entire ocean community. Researchers say the trillions of *Prochlorococcus* in the seas weigh as much as 220 million Volkswagen Beetles, providing food to thousands of other ocean creatures and generating 5 per cent of Earth's oxygen.

Scientists theorise that *Prochlorococcus* once had the gene to produce the enzyme, but that living in a cooperative community where other cyanobacteria do the job for free allowed it to lose that gene. This theory is hilariously called the Black Queen Hypothesis, after the card game Hearts, in which players ditch cards trying to avoid getting stuck with the queen of spades. It counters another biological theory called the Red Queen Hypothesis, which suggests that competing organisms are involved in a constant evolutionary arms race — based on Wonderland's Red Queen telling Alice that 'it takes all the running you can do, to keep in the same place.' Most of us assume that evolution tweaks organisms to become more complex in order to succeed, but scientists theorise that evolution works in the reverse way, too: when organisms obtain shared goods from others in their community, they can afford to become less complex. Reminds me of some marriages: he forgets how to do the laundry, she forgets how to fire up the grill, and they use the freedom granted by their union for something else. Gardening? Reading? Petting the dog?

We humans have prospered as a species for a number of reasons, but one is certainly the marvellous inventions that have allowed us to relax selection. Agriculture and medicine, buildings and heating systems, and traffic signals and bike helmets — all allow us to thrive. Our problem now is that many of these inventions and our resulting abundance increases the pressure on the rest of nature, rupturing ecosystems and driving other species to extinction. And the abundance itself doesn't even nurture everyone within our own species, with some so wealthy they have gold toilets and others so poor they squat in ditches. We need to relax our pressure on the rest of nature, guided by new metaphors for our relationships with other species — metaphors that will hopefully spill over into our relationships with each other. We need to be sweeter.

# Chapter 3

# We Are Ecosystems

One of aviator Amelia Earhart's last acts before her disappearance in 1937 was a joke played in microbes.

She and navigator Fred Noonan were just beginning their historic circumnavigation of the globe — not the world's first attempt, but one that would, at 29,000 miles, be the longest — and planned to help scientists back home by snatching up air samples that might contain the fragments of life as they careened through the atmosphere. The US Department of Agriculture had equipped their Lockheed Electra with a 'sky hook', a device that looked something like a piccolo, created by phytopathologist Fred Meier and used by Charles Lindbergh and his wife, Anne Morrow Lindbergh, in their 1933 flight across the North Atlantic Ocean. As particles flowed through the sky-hook tube, they were captured on a slide inside an aluminium cylinder for later study. While flying, Earhart and Noonan would manually open the tube for thirty minutes. Then, Earhart wrote, 'the slide within it is exposed to the moving air and gathers upon it whatever minute beasties may inhabit the particular stretch of atmosphere just then being flown through ... Subsequently, the cylinder was closed, sealed, and the place and time of its exposure recorded.'

They made around a dozen of these recordings in their flight from Miami to Africa, but accidentally made the first before they even took off. They were fiddling around with the equipment, trying to get the hang of how to handle the cylinders, when Noonan coughed on one of the slides. 'That's ruined,' said Noonan, according to Earhart's account, and he reached out to throw it away. 'The collection of germs on that slide would look like a menagerie under a microscope.'

But Earhart insisted on keeping that slide and cylinder among the specimens that they'd later hand over to the scientists. 'I thought it would give the laboratory workers something unique to ponder when they came upon its contents among the more innocent bacteria of the equatorial upper airs,' she wrote. 'Heaven knows what cosmic conclusions Fred's contribution might help them reach!'

But surely Noonan's sneeze couldn't have contained stranger organisms than the ones already afloat in the atmosphere, out of sight and range of the formerly earthbound.

Bacteria (single-celled organisms with no nuclei) and protozoa (single-celled organisms with nuclei that belong neither to the world of plants, animals, nor fungi) were first spotted in the late 1600s by Antonie van Leeuwenhoek, a Dutch fabric merchant and haberdasher who unexpectedly became one of his era's greatest scientists. Living things too small for the human eye to see had been newly revealed by microscopes, and drawings of them were all the rage — in fact, a book called

*Micrographia* was 'practically a fashion accessory' in London, according to British biochemist Nick Lane. Van Leeuwenhoek determined to grind his own lenses to create a microscope that peered even more deeply into the invisible world. He succeeded in creating the most powerful microscope of his day and turned his curious eye on everything around him. He wrote letters about his discoveries, which were published in the British journal *Philosophical Transactions of the Royal Society*, and first mentioned seeing microorganisms in a 1674 letter. He described streaks of summertime algae in a nearby lake, and then penned these Earth-shattering lines: 'among these streaks there were besides very many little animalcules ... And the motion of most of these animalcules in the water was so swift, and so various upwards, downwards, and round about that 'twas wonderful to see: and I judged that some of these little creatures were above a thousand times smaller than the smallest ones I have ever yet seen.'

Because of his mercantile background and the fact that he could see things with his microscopes that no one else could, van Leeuwenhoek's claim proved controversial for a while. Eventually, though, other astute observers also began to detect microorganisms. These revelations weren't regarded as having much bearing on human concerns until 1876, when German microbiologist Robert Koch announced his discovery that one of the animalcules, *Bacillus anthracis*, was connected to a specific disease — anthrax in cattle — and later figured out how to grow bacteria in agar cultures in a lab. At around the same time, French chemist Louis Pasteur discovered the connection of microorganisms to fermentation. Scientists of his time largely believed that fermentation was a merely chemical process, but Pasteur proved in his lab that it can only occur with the presence of living microorganisms and that the latter could arrive by air.

Pasteur was fascinated by the presence of airborne microorganisms and roamed the countryside collecting samples in flasks, convinced that living bacteria and mould spores are not only drifting in the air but vary in number depending on the location. He wanted samples from higher elevations and considered conducting experiments in a hot-air balloon, but settled for climbing mountains and snagging samples at 850 and 2,000 metres above sea level. Aviation pioneers of the early twentieth century were finally able to sample the air thousands of feet higher, giving a huge boost to scientists who had been pecking away at the ubiquity of microorganisms on our planet for some 250 years.

Scientists are still eager to see what kinds of microorganisms are blowing around the atmosphere and what effect they're having, only now the scientists are often up in the planes themselves. Athanasios Nenes, a professor of atmospheric processes in Switzerland, has flown above hurricanes — up to around 33,000 feet, way out at the edge of the troposphere — to look at the impact these massive weather disturbances have on the environment. One of the things he and his colleagues did was to collect samples of the particles blowing around up there. Hurricanes move like enormous vacuum cleaners over the land, sucking up particles and expelling them into the sky, where trade winds pick them up and can waft them around the globe in a week. The team was surprised by the number of living microbes among the particles, even under such formidable circumstances.

'They weren't killed, even though they were exposed to low temperatures and dry conditions and zapped with ultraviolet radiation,' Nenes says. 'When we got them back to the lab, most of them grew and were pretty happy.'

Whenever scientists probe such extreme environments — boiling geothermal vents, stinking sulfuric ponds, Antarctic ice, the bottom of the ocean with tons of water pressure bearing down — they find bacteria. In fact, wherever scientists focus better tools on areas that they assumed had no life, they are astonished by an abundance of life. In late 2018, a group announced the discovery of a vast underground ecosystem containing an estimated twenty-three billion tons of microbes — weighing hundreds of times the collected weight of humanity — three miles under the Earth's surface. They survive in extreme heat, with no light, under intense pressure, and without much to eat. Some breathe uranium. Some seem to live in this hot, dark world indefinitely, in a sort of suspended animation. 'The strangest thing for me is that some organisms can exist for millennia,' University of Tennessee scientist Karen Lloyd told the *Guardian*. 'They are metabolically active but in stasis, with less energy than we thought possible of supporting life.'

The study of microorganisms and their impact — in every environment as well as inside the creatures living in those environments — is one of the hottest areas of science today. It's also one of the hottest topics of conversation among people trying to figure out how to be healthy. You probably already know that, even just from wandering through a grocery store, because the makers of prepared foods and other products are trying to turn the science into a marketing strategy. There are now 'probiotic' — supposedly containing beneficial bacteria — bottled waters and chocolate bars in the United States, probiotic pastas in Iran, probiotic muesli in India, probiotic pancake mix in Puerto Rico, and probiotic toothpaste and mouthwash in China. But despite what might be massive and dismaying overhype by manufacturers, it is an amazing time to be alive and to reconceive our connection to the rest of life on this planet.

Most of what we believe about the behaviour of living things and their impact on ecosystems is based on thousands of years of observation of macrobiology — plants, fungi, and animals — but we now understand that all of us macros are like meat and vegetable floaters in an incredibly vast broth of microorganisms. Or like stars and planets suspended in the overwhelming vastness of dark matter. We are swathed to each other and the rest of the natural world by an invisible cloud of bacteria and other living things — including things so tiny and weird that they make scientists argue about the definition of the word 'alive'. These microorganisms are critical to the functioning of just about every habitat — from that of a mountain stream to the ones in our tear ducts — and it's become clear they must be part of our every consideration, whether we're building a hospital or replanting a forest or simply making dinner. The new understanding of their presence and impact is driving scientists to reconsider everything they think they understand about the natural world, which, of course, includes us.

Ten years after Robert Koch proclaimed the connection between bacteria and anthrax in 1876, Dutch microbiologist Martinus Beijerinck showed that bacteria could be beneficial as well as harmful. He explained a phenomenon that farmers had observed for millennia, namely that planting legumes like peas or clover restored fertility to fields heavily planted with wheat and other crops. Beijerinck found that inside nodules on the roots of the legumes were bacteria that converted ('fixed') atmospheric nitrogen into a plant-available form, thus fertilising the fields. Despite Beijerinck's contribution, though, the continuing view among scientists and laypeople alike was that bacteria and other microorganisms were a lethal if invisible force, something to make war on chemically or to avoid.

As late as the 1970s, science was largely clueless as to the role, volume, and astonishing diversity of the microscopic life in and around us. Biologists traditionally traced the evolutionary lineage and genetic relatedness of plants, animals, and fungi by studying their physical features and fossil records, and had concurred that there were thirty to forty different major groupings, or phyla, of animals. But it was nearly impossible to classify bacteria and other tiny organisms this way.

'Microbes don't have a lot of morphologies,' says Margaret McFall-Ngai, a biologist who studies the mutualism between the Hawaiian bobtail squid and a bacterium that grants the squid luminescence — with this superpower, the squid creates a special light organ that allows it to blend into the moonlight shining over a dark sea, erasing its silhouette so it can sneak up more easily on prey and avoid predators. The study of mutualisms like this were considered oddball, backwater science back when she got started some forty years ago, because the prevailing view was that there weren't many of them — the opposite of what we are now discovering. 'There are only about five or six morphologies that you can detect under a microscope. And we don't know how to culture most bacteria [i.e. grow them in a lab], so it was hard to study them. It was a big mystery world out there,' says McFall-Ngai.

Then, in the late 1970s, Carl Woese revolutionised biology by introducing a new way to classify microorganisms by sequencing a slowly evolving gene that all living organisms share. This, McFall-Ngai says, gave microbiology something it had never had: evolutionary sweep, an understanding of relatedness among the microbes and of the way they have changed over time. At first, sequencing was expensive and laborious, but both the price and the time required dropped precipitously over the years. As the technology improved, estimations of the number of bacterial species began to spike, and the upward sweep continues today. In 1990, Woese published a paper suggesting that there were dozens of bacterial divisions — the major lineages — not four, as was believed in the 1970s. More recently, in 2014, Pablo Yarza and colleagues published a paper in *Nature Reviews* suggesting that the number of bacterial divisions could be as high as 1,350.

'Fifty years ago, we thought there were thirty to forty divisions of animals, and today — depending on whose organisation you're adhering to — we still think there are thirty to forty different kinds of animals,' says McFall-Ngai. 'But bacteria have gone from four to 1,350. And the thing is, they run everything.'

The latest estimate is that there are a trillion different bacterial species in the world, and that they've been on the planet for some 3.8 billion years. Given their vastness and diversity, none of us macros would be around if a significant proportion of them were harmful to us. 'The proportion of all bacterial species that is pathogenic to plants and animals is surely small,' University of Hawai'i biologist and professor emeritus Michael Hadfield told *PhysOrg* in 2013. 'I suspect that the proportion that is beneficial/necessary is likewise small relative to the total number of bacteria present in the universe, and surely most bacteria, in this perspective, are "neutral". However, I am also convinced that the number of beneficial microbes, even very necessary microbes, is much, much greater than the number of pathogens.' So most bacteria are indifferent to us and are just carrying on their own business — and often, their business is running the ecosystems on which we depend. Some are dangerous, but many more are active cooperators. Without them, life as we know it would not exist.

Bacteria are not only numerous and diverse, but also highly adaptable. They reproduce quickly — some in only twenty minutes — but unlike us, they don't need to mate and reproduce to gain greater genetic diversity. Instead, bacteria living in proximity to one another — often in large, protective colonies or simply jammed together into a space cheek by jowl, as they are in our guts — can simply trade with their neighbours for the DNA they need to handle a specific problem. I'm a brown-eyed woman who married a blue-eyed man and we produced a lovely blue-eyed daughter. But if I were a bacterium needing the equivalent of blue eyes, I could just get that DNA from a neighbour without sex and reproduction — it's called horizontal gene transfer. In fact, that was the first sex. As Lynn Margulis and her son Dorion Sagan wrote in their book *Microcosmos*, 'Sex, as recognised by biologists, is the mixing or union of genes from separate sources. It is not to be equated with reproduction … Bacteria sex preceded animal sex by at least 2,000 million years and, like a trump card, it permitted all sorts of microbes to stay in the evolutionary game.'

Here's a puckish take on one way that they stayed in the evolutionary game — and on the relationship among us visible organisms and the invisible microbes: 'I see plants as microbes' first experiment with artificial intelligence,' says New Mexico

State University molecular biologist David Johnson. 'They needed something else to capture energy for them, so they helped develop the plant. Then they realised plants weren't mobile enough to get them all the food they wanted, so they developed version 2.1, animals. Somewhere along the way, they went horribly awry and came up with humans.'

Not only do scientists now believe Lynn Margulis's hypothesis that bacteria and other single-celled organisms cooperated to form the eukaryotic cells that went on to become the domain of life — including plants, fungi, and animals — many also believe that much of the continuing evolution within the plant, fungi, and animal kingdoms is driven by microbes. In this view, we eukaryotes didn't evolve solely based on the traits and abilities encoded in our genomes. We evolved through a complex interaction among us, the environment, and our microbiota (though many people use these words interchangeably, microbiota refers to the community of microbes, and microbiome refers to both the microbes and their collected genomes). In fact, scientists believe we have an evolved dependence on our beneficial microbes, which British microbiologist Graham Rook and other scientists call our 'old friends': we can't get along properly without them. And just as microbes have pushed their hosts' evolution in certain ways, their hosts have pushed the microbes that specialise in them. Sometimes the microbes evolve so much to suit their hosts that they can't live anywhere else. For instance, scientists estimate that 90 per cent of the bacteria in termite guts are not found elsewhere.

Basic aspects of human physiology can be explained by the importance of our partnership with microbes. For instance, when we're sick, we take out a thermometer and check to see that our temperature hovers around our normal 100 degrees Fahrenheit (40 °C). Some researchers think we evolved endothermy — the ability to maintain that temperature — because it's the temperature at which our microbiota operate most efficiently. Our immune system may have a similar raison d'être. We are only fully aware of this system when we're sick and it battles what it perceives as invading pathogens with fevers, chills, inflammation, sneezing, snotting, and more. Accordingly, we have always believed that our immune system developed to protect us from germs, but scientists — with McFall-Ngai leading the thinking on this with a 2007 essay in *Nature* — now believe the immune system's primary role is the daily maintenance of our immense microbiota. As we take microorganisms and other pieces of the wider environment into our bodies — not just conveyed by what we eat or put into our mouths by biting the random fingernail, but also by what we breathe and swallow — the immune system samples these particles as they accumulate in our small bowel and determines which microorganisms should be recruited and welcomed into the fold and which to guard against.

Scientists now find that just about every complex plant, fungus, and animal hosts a dynamic microbiota — a community of bacteria, fungi, viruses, protozoa, and other microorganisms that live in us and on us and are essential to our health, just as we are essential to theirs. We are not individuals but ecosystems, each of us hosting a whirl of organisms busily interacting with us and with each other in a complex web of connection. The ecosystem of us lives within larger ecosystems — our gardens, our neighbourhoods, our farms, our cities, whatever wilderness we have left — where we interact and overlap with other animal-plant-fungal ecosystems and the invisibles inside them as well as free-living microbes throughout the environment that aren't associated with hosts. Those larger ecosystems are nested within and interacting with even larger ones, on and on until they encompass the entire planet.

And if the rub-your-stomach, pat-your-head mental exercise of picturing those nested ecosystems isn't mind-boggling enough, consider that even our microbiota have a microbiota. Viruses and even smaller mobile 'genetic elements' (meaning, things that move bits of genetic material around but don't meet the traditional definition of life) weave in and among the bacteria in our guts, changing the affected bacteria's own genetic potential for harm or good. A new field called infection genomics is studying this microbiota of our microbiota, but already scientists have gotten a few glimpses of how it operates. One example: when we take broad-spectrum antibiotics, that action indiscriminately stresses and kills many of the bacteria in our guts. This can trigger infectious elements that confer antibiotic resistance to move along little bridges from one bacterial cell to another, ultimately spreading antibiotic resistance throughout our entire system. 'These infectious elements have their own evolutionary pressures — sometimes it's good for them to move and sometimes it's best for them to stay with their host,' says University of Illinois microbiologist Rachel Whitaker. 'When they sense that the cell is stressed, they move and take antibiotic resistance to protect the next host.'

In this case, the infectious elements are protecting their hosts — the bacteria — and thereby making our microbiota resistant to antibiotics, which may ultimately cause problems for the bacteria's host — us. But for the most part, most viruses are as harmless to us as most bacteria, although they share the same bad rap — a bad rap that begins with their very name, as 'virus' is derived from the Latin word for poison. 'Twenty years ago, people thought all bacteria make you sick, but hardly anyone thinks that now because we now understand how important bacteria are,' says Pennsylvania State University professor of virus ecology Marilyn Roossinck, the author of *Virus: An Illustrated Guide to 101 Incredible Microbes*. 'I think people will hopefully get to that place with viruses, too. We need them! Fewer than one per cent of them give us trouble.'

Like bacteria, viruses are incredibly ubiquitous — and even greater in numbers — and very active in our world. Scientists recently calculated that every day, some 800 million viruses fall on every square metre of the planet, swept up like the bacteria in Nenes's survey and blown around the world to fall down again like an invisible, steady rain. Like bacteria and other microorganisms, viruses can be symbiotic (meaning, they live in close proximity to their hosts), infecting their hosts with their DNA as a way of replicating, as they don't have any reproduction machinery of their own. For that reason, some scientists don't consider them living things, although Roossinck isn't buying that. 'They borrow the machinery of the host,' she says. 'I just think that means they're clever, not that they're not alive. They're alive when they're inside a cell and dormant when they're not. I think they were probably the original life form.'

Those infections are usually not harmful, and sometimes they offer the host a fabulous gizmo from the vast closet of viral DNA that changes the host's genome forever. In fact, scientists think that 40–80 per cent of our genome may be connected to long-ago viral infections. An ancient retrovirus infected one of our four-legged ancestors and left behind a patch of DNA that is now part of our nervous system and critical to consciousness, memory formation, and higher-order thinking. Another viral contribution Roossinck likes to point out is the mammalian placenta, which is created when a protein fuses a group of cells together. That protein gets its marching orders from a retrovirus that became part of the mammalian genome 160 million years ago.

Roossinck became fascinated with viral symbiosis when she and colleagues conducted a study in Yellowstone National Park ten years ago. They found that some plants could tolerate the extreme heat near geysers and hot vents, but only if the plants had a symbiotic mutualism with a certain fungus that was infected by a certain virus. 'It was a beautiful three-way symbiosis,' she says. 'I think they're very common, but it's hard to study them.'

In fact, scientists have been finding that some of the most well-known symbiotic mutualisms are more complicated than they thought. Take that classic mutualism between leguminous plants and bacteria in which plant roots form nodules to house the bacteria and, in exchange for food and housing, the bacteria convert nitrogen into a plant-accessible form. Turns out that symbiosis can only take place when a certain plasmid — a circular molecule of DNA that's separate from a cell's chromosomal DNA — is present in the bacteria. It may be that all the important work that bacteria do in the environment — as well as within the organisms that live there — is enabled by some sort of mutualism with virus-like things and other microorganisms we haven't yet identified.

For most of human history, we have been dazzled by the miracle that is us — our ideas, our innovations, our inventions. Some Indigenous cultures understood how deeply human fate is tied to respecting and supporting the rest of nature, but the historically dominant, conquering viewpoint sees humans as the pinnacle of evolution and the masters of the planet. But our new understanding of the invisible world of microorganisms changes that: we are not on top of the heap but are, rather, part of the heap. Everything is part of the heap. And the heap is in us.

We can't separate ourselves from the rest of nature —

and if we try to do so with our innovations and inventions, we risk diminishing ourselves. Because nature is not just the stuff outside the boundaries of our cities or protected in a distant park or growing in an abandoned lot or blown in on a summer storm, like some tropical bird that city residents in the equatorial north discover pecking at cracks in the pavement. Nature is also our human parts plus the societies of microorganisms thriving all over our bodies, and, most critically, the multitudes hunkered down in and nibbling at the mucus lining our guts and influencing or regulating everything from our metabolism to our mental processes. Fifteen per cent of the small molecules in our bloodstreams come from these gut microbes, and those molecules interact with every cell in our body.

If we could somehow vaporise all our human parts, we'd leave a microbial outline. We think we're eradicating germs when we use hand sanitisers or even wash our hands vigorously, but our own symbionts — called germs until recently — retake their position in seconds.

'All you're doing is wiping off the patina, the stuff you just encountered in the last few minutes,' says Jack Gilbert, a professor of both paediatrics and oceanography at the University of California, San Diego. 'Your skin is a deep three-dimensional tissue, sort of like Manhattan. The stuff you wipe off is at the top of the skyscrapers, but the actual microbial world is deep down in the streets, inside the buildings. Your microbes will rapidly recolonise the tops of the skyscrapers.'

And a good thing that is, too: the skin microbiome has been shown to help with wound healing and protection against pathogens.

Research into the world of microorganisms falls into two big clumps: studying the ones with hosts and studying the free-living ones throughout the environment. In the real world of microorganisms, though, the line is blurred. Like the *Peanuts* character Pigpen, we constantly spray our resident microbes around whatever environment we happen to be in. According to Gilbert, humans emit thirty-eight million bacteria and seven million fungi in an hour. 'They're shedding from our skin, from our hair, every time we breathe out,' he says. 'If you're sitting in a chair, they're coming out your rear end. We leave a microbial signature.'

My dear departed mother would have squawked at this news and rushed to buy antimicrobial sprays and wipes. And back in the days when microbes were universally shunned, even scientists couldn't understand why hugging, kissing, hand-holding, and other physical intimacies persisted in so many cultures. Didn't these behaviours spread germs and make people sick?

But it seems that nature intended that we share our microbiota. According to research by evolutionary biologist Andrew Moeller of Cornell University, chimpanzees who engage in lots of these intimate social behaviours had a more diverse microbiota. 'Now that we know that most microorganisms are commensal [just living there peacefully] or even beneficial, it seems reasonable to think that this transfer of microbes is in many ways beneficial,' he says.

Our personal microbiota comprises some thirty-five trillion bacteria — we have around the same number of human cells as bacterial ones — as well as fungi, protozoa, viruses, and other microorganisms. We have dozens if not hundreds of habitats — in our gut, but also in our mouth, our lungs, our eyes, our armpits, our urogenital tract, on our skin — where bacteria and other microorganisms participate in the processes that keep those areas healthy. 'I think of our bodies as a world with all these different habitats where microbes live,' says Arizona State University evolutionary biologist Athena Aktipis. 'As we go about our daily activities, they can move from one region to another.' The densest and most significant part of our microbiota is in our gut, where billions of our tiny symbionts range along the 344-square-foot surface area of our gut and participate in processes that are essential to the body's overall health. Scientists believe that the assemblage of our microbiota begins as we pass through the birth canal and emerge with a life-giving slick of microbes from our mother's vagina and, despite the best efforts of the nursing staff, a smudge of faecal bacteria.

'Most species poo on their babies as they're being born,' says microbiologist Graham Rook. 'Koala babies lean out of the pouch and eat a special form of maternal poo. Passing the maternal microbiota to the child is immensely important.'

The microbiota of breast-fed babies gets an immediate

boost after birth, as breast milk is loaded with carbohydrates called oligosaccharides that we humans can't digest, but which our microbes love — it seems that one of a human mother's first tasks is to nourish her child's starter kit of microbes. That starter microbiota grows in diversity as the baby interacts with the outside world. As soon as they're old enough, babies put everything in their mouths, from their toes to dead flies to the contents of a litter box if they get the chance; according to one group of researchers, they'd eat twenty grams of soil a day if adults didn't keep swooping in to interfere. It seems that this is part of a biological strategy to build a robust microbiota and immune system. By the age of three, the child has a stable signature microbiota as large and diverse as that of an adult.

But that doesn't mean that it's completely static. Scientists studying the microbiota find that it is dynamic; depending on the environment, the ratio and proportion of its members shift. For one thing, the various populations in our gut rise and fall in numbers, depending on what and when we're feeding them. They may be less numerous in the mornings before we've eaten. Then, their numbers grow as our food selectively fuels the growth of whatever microorganism prefers it. When we eat lots of plant material, for instance, that 'fibre' — which mostly means plant cell walls — is set upon by the many bacteria that have the ability to break down complex carbohydrates, and they grow in number. And it's great to have lots of them, as they convert that fibre, which we cannot digest, into the short-chain fatty acids that are crucial to our overall health. They also exude byproducts that feed other gut microbes. 'When you consume more complex carbohydrates, you are actually taking care of all these different food chain levels,' says Athena Aktipis. 'That creates a more diverse and resilient ecological system inside your body.'

When we eat sugary junk food, sugar-eating microbes proliferate, too, but their abundance doesn't confer the same kind of benefits — quite the opposite. According to a paper by Aktipis, sugary food actually sets off a fight among certain microbes and our body's own cells, both of which are eager to grab up the easy energy afforded by simple sugars. Adaptability seems to be one of life's bedrock characteristics: these microbes flock to simple sugars because they yield energy easily, especially compared to the biochemical locks and bolts involved in wresting energy from complex carbohydrates. The sugar-eating microbes that proliferate by eating sugar are often pathogenic and, a recent study in mice has shown, may also prevent beneficial microbes from colonising the gut. Soon, the body responds with inflammation to beat them — but the sugar-eating microbes can actually become more virulent as they fight to monopolise the simple sugars.

Scientists think a healthy microbiota is tied to our every organ system, function, and trait. 'That's not surprising, because the microbiota are so central to our metabolism,' says Andrew Moeller. 'And the metabolism is central to every tissue and organ system in the body. Metabolism is basically how animals change things from the environment into more animals.' On the other hand, a slew of physical ills are tied to disturbances and lack of diversity in the gut microbiota, including allergies, autoimmune diseases, obesity, diabetes, cardiovascular disorders, cancer, and central nervous system dysfunctions like learning and memory impairment, anxiety, depression, and autism.

What causes these disturbances in our microbiota? One way to characterise the problem is that it is another 'dis' — a disconnection from the natural world as the mammals that evolved into humans, and even our much more recent human ancestors, experienced it. A disconnection from the full panoply of microbial life, both that which resides inside us and the living richness outside us.

Babies born by caesarean section have a diminished starter microbiota, although some enlightened doctors try to help them catch up by smearing them with fluids from their mother's vagina. And the microbiota of babies who aren't breastfed, even for a short period, miss out on the special nutrition that nature designed for them.

We can further damage and diminish our microbiota with antibiotics, which are chemicals based on ones produced by bacteria in nature. Some scientists believe that in their natural state, these chemicals are not the weapons of mass bacterial destruction that they become in our highly concentrated medications; rather, they are instruments of communication, ways for bacteria to tell each other to back off or to stimulate some other behaviour from a neighbouring bacterium or colony. 'We have evidence from lab experiments that low concentrations of antibiotics turn on certain genes in other bacteria,' says microbiologist Julian Davies of the University of British Columbia. 'Seems to me the role of antibiotics as signalling molecules makes sense. It's like a telephone: it gives information, but doesn't kill you.' There are times when we need the powerful antibiotic compounds that are in our medications, but doctors are much more cautious about prescribing them than they used to be.

And, of course, there is the daily impact of the modern world's version of food and shelter. The great majority of our food is raised in outdoor factory farms in which the natural partnerships between microorganisms, plants, fungi, and animals have been destroyed (more on this in chapter five), and the managers of these factories try to replace the benefits of these partnerships — like fertilisation and resilience to pests and disease — with synthetic chemicals. These debased foods then enter the vast maw of mass-market manufacturing, which processes them and mixes them with more synthetic chemicals, rids them of any microbial life — they might spoil! — and burps them out in one packaged product after another. It's hard to think of these products as actual food. As Michael Pollan says, they are merely 'edible food-like substances'. When we attempt to live on this super-processed stuff so far removed from nature — when we give up fresh fruits and vegetables and animal products from good farms or forage — we diminish the diversity of our microbiota with every breakfast, lunch, dinner, and snack.

The average human in industrial societies spends around 90 per cent of their time indoors, which further impoverishes their microbial community. We may fill up rooms with plumes of our own microbiota, but modern buildings have been designed and engineered to keep the rest of nature out. We keep our windows tightly closed — often, even in the summertime — and breathe air that has been carefully filtered to keep dust, pollen, and microbes out. If we have dogs, they pad and shed some of that wilder nature back inside — and scientists have noted their positive impact on the microbiota — but aside from that, we spend most of our lives in sterile, microbially bleak surroundings.

Studies show that the combined effect of all these modern innovations has reduced the healthful diversity of our microbiota. One compared the microbiota of children in a Burkina Faso village where the local diet is similar to that of humans around the time of the birth of agriculture — lots of plant material — to that of children in rural Europe. The African children had not only greater diversity in the microbiota, but also significantly more healthful short-chain fatty acids than the European children. Other studies that compare urban and rural populations in Europe — especially rural populations in which people get dirty working with pigs and cattle and other animals — find a similar disparity in their microbiomes.

Studies in animals show that those of us living on a debased Western diet of poorly grown, highly processed food — and especially those of us who were dispatched into this world with a poor starter kit of microbes and were then dosed with too many antibiotics at some point — can develop such an impoverished microbiota that key players go extinct within us. Meaning that no matter how many fibrous fruits or vegetables we eat to make up for our past, we can't regrow certain essential microbes if they've been wiped out.

But some scientists think we may be able to replenish even the most impoverished microbiota when we reconnect with nature.

Scientists have recently been questioning the stability of the microbiota, and now suspect that it is far less rigidly fixed than previously assumed. It appears that some microbiota comprise stable members as well as transient members that don't colonise but just pass through. 'This is the nuance the field is starting to appreciate,' says Vanderbilt University biologist Seth Bordenstein, pointing to research from 2013 that suggested that only 60 per cent of the human microbiota was stable over five years. 'We don't yet completely understand how constant the ecosystem is through time and geography.'

Some studies suggest that at least some of the turnover may be of spore-forming bacteria, which can leave the human body through our faeces and then survive externally as microbial seeds for hundreds and even thousands of years. Anywhere humans have lived, the soil is seeded with human-adapted bacterial spores from previous generations. Some scientists believe that we modern humans can reseed ourselves with these bacteria by immersing ourselves in nature and, especially, spending some time around good soil. 'I mean, I'm sure I have organisms in my gut derived from Julius Caesar's poo,' quips Graham Rook. 'He would have passed through London. He would have pooed around here. The spores are numerous and resilient, and they must still be out there.'

At least one study has shown that city folks' microbiota can change when exposed to more traditional food and lifestyle. In this case, five urban adults and two children stayed in a Venezuelan rainforest village for sixteen days. They ate the local diet of cassava, fruit, and fish with the occasional bite of game meat and followed the village lifestyle in other ways, bathing in the rivers without soap or shampoo and eschewing toothpaste. All the surveyed areas of the city folks' microbiota became more diverse, including the gut, skin, mouth, and nose, with significant increases in the children. Another study from Finland suggests that we can bring the benefits of nature into the urban environment: when schoolchildren were exposed to debris from a forest floor, their blood showed an increase in the biomarkers of a robust immune system.

All of this new microbiota research echoes older epidemiological studies as well as folk wisdom, which suggest people living closer to natural areas are healthier. Those studies target long-term health benefits, including lower death rates, reduced cardiovascular disease, fewer allergies, and reduced psychiatric problems. People have assumed that these health benefits were largely psychological — that the experience of

forests and meadows delivered a salubrious mind-body jolt
— or were the result of the increased exercise or exposure to
sunlight one often has when enjoying nature. And of course,
those are good, healthful things. But, as Rook argued in a paper,
these effects are fleeting and not necessarily tied to long-term
health. Rather, what might explain the healthfulness of natural
areas is that we encounter the more diverse microbial life that
partnered with our ancestors long ago. The illnesses that plague
Western society are largely connected to malfunctioning
immune systems and inflammation, and these, in turn, are
connected to our impoverished microbiota.

Rook does a bit of public speaking about his microbiota
research, and in one session a pig farmer piped up to say that
if one of his sows was raising piglets in a concrete sty, he'd
be sure to throw in a clod of turf to keep them healthy. 'It's
been shown that pigs brought up in a field are healthier and
have less inflammation,' Rook says. 'Same is true in a lab with
germ-free mice [which are born by C-section into a sterile
environment]. If you put them on soil instead of clean bedding,
they have fewer allergies and will be healthier. In fact, you can
reconstitute their guts with soil microorganisms and they'll be
okay. If you then expose them to mouse-adapted strains, those
will displace the soil microorganisms quickly. Nevertheless, it
shows how contact with the environment is crucial.'

So we might venture out into the rest of nature and wind
up with a bacterial spore deposited by Julius Caesar or George
Washington or Sojourner Truth or John Muir that stakes a

new and valuable claim in our gut. Or we might encounter microorganisms that don't colonise but are like a beloved guest that passes through and leaves gifts in her wake. One of these is the bacterium *Mycobacterium vaccae*, so named because it was commonly found in areas where cows have been kept. Scientists were testing whether this bacterium might be helpful for cancer therapy. It didn't beat back the cancer, but the patients who received the treatment showed unexpected psychiatric improvements. So a group of scientists did a mouse study in which mice stressed by a dominant male — usually, they'd develop a stress-driven colitis from such an encounter — were immunised with a heat-killed version of the bacterium. They found it reduced the mice's level of stress and stress-related inflammation. 'The animals become resilient not only to the psychological and brain chemistry effects of stress, but also to inflammatory disease in the bowel,' says Rook, who was an author of the study. 'It's a pretty exciting paper. And this is the kind of organism you'd find just mucking around in your garden.'

This is why I keep asking the young woman who sells me grass-fed beef if she'd please bring me a bag of cow patties, too. I want to feed my garden.

There are quite possibly thousands more microbes found in the soil and on trees and on mossy rocks and clinging to tide-swept shores that have similarly beneficial impacts as they make their way through our bodies — scientists couldn't begin to study them all, but we can assume they're there. They might not join our resident colony of microbes, but they benefit our bodies in yet another important way: they continue training our immune system to distinguish between good and bad actors. We may have spent our babyhood sucking on dog ears and eating garden snails (I can still remember the crunch) in an unconscious gambit to build a healthy immune system, but the system needs reminders. Re-education helps ensure that our immune system continues to fight the bad guys and — just as important — not battle innocuous microbes, pollen, and other harmless stuff.

The disconnection from the rest of nature could partly explain why elderly people falter when they move into nursing homes and other care facilities. We assume the decline is purely psychological — from the loss of independence and familiar surroundings — but part of the decline might be caused by the physiological toll of isolation. One study showed that people in long-term care had less diverse microbiota as well as higher levels of inflammation and poorer health. 'It seems there is a requirement of continuing input to maintain the biodiversity of the microbiota,' says Rook.

There are many reasons for humans to respect and safeguard nature, from the wildest parts in remote protected areas to the less wild nature in our backyards. The people reading this book know those reasons. But certainly, the crucial dialogue and mingling between our microbial partners and the microbes in the greater environment outside us is another. Can we deplete and damage landscapes so much that the dialogue stops, not just for us but for our fellow macros in the animal, plant, and fungal kingdoms?

And how much damage can the many nested macrobiota of the world sustain before they stop functioning? If microbes speak to each other in chemicals, then what are the chemicals coming out of our smokestacks and drainpipes and crop dusters telling them? How are the fragrances from my neighbour's dryer vent and the combustive stink from Fourth of July fireworks constraining them? What orders or disorders are barked at my skin microbiota by the transformative stench my hairdresser uses to turn my hair from grey to brown? That, not the cost, is what made me give it up.

# Chapter 4

# Transforming Deserts into Wetlands

Carol Evans and I push aside the remains of our vegetarian pizza and white wine and lean toward her open laptop. We both peer at the Dalí-esque image of an empty metal ladder tilting in the middle of a blue creek fringed by green and gold rushes. Greyish shrubs loom just past the rushes and then dwindle into the sere yellow hills behind, which appear as rumpled as a tossed blanket. The ladder looks as if it's perched on top of its own reflection in the water below and about to be rocked by a ring of waves heading toward it.

'I tried to retake the photo from the same exact spot on Maggie Creek where the original was taken in 1980,' explains Evans with a laugh. She is a recently retired stream biologist with the Bureau of Land Management (BLM) in Elko, Nevada, and I had hoped to go out in the field with her on this trip. I wimped out when her colleagues warned that Evans has such gusto for this work that she can't help tramping the back country for miles and miles when she goes out — drawn ever farther by the possibilities beyond this stand of willows or that hill — and that I might have a hard time keeping up.

'People could walk all the way out there back then, but it's covered with water now,' Evans continues, pointing to the ladder. 'I got up on the ladder to try taking the photo, but it kept sinking.'

The original 1980 photo had been taken as part of an inter-agency effort beginning in the late 1970s to document the alarming deterioration of area streams that was taking a toll on the native Lahontan cutthroat trout. The photo was one of thousands taken along about 1,000 miles and more than 150 streams in the Elko District, including Maggie Creek and Susie Creek watersheds, in addition to measurements of factors like stream depth, width, temperature, and the amount of vegetation on the banks and their stability. Evans has been working in and around these streams for over thirty years, and was in the process of examining and re-photographing the original sites in the Maggie Creek and Susie Creek watersheds to document the great change that's taking place. She herself had been the source of the waves cruising toward the ladder as she slogged around the creek in her waders a few weeks before we met in early October 2017, trying to get as close to the original spot as she could without going under.

Aside from the ladder, this was a beauty shot, with an abundance of water and willows, sedges, cattails, and other riparian vegetation hugging the creek banks. A spine of high-desert hills arched in the background. Then Evans flipped through the hundreds of photo files on her laptop to show me the photo from thirty-seven years ago. In October 1980, this site — the Maggie Creek Stream Survey Station S-9, just north of Carlin, Nevada — was all about desert. The only vegetation in the photo was sagebrush and rabbitbrush, upland shrubs that were far from the stream and laden with dust. There was no riparian vegetation at all, and the ground surrounding the narrow creek was as desiccated and yellow as the distant hills. In this older photo, a moustachioed young man stood in the exact same spot as Evans's ladder — back then it was dry land — holding out a large card with information about the site's location, with a mark indicating that the photographer was looking upstream. This was before GPS and geo-referenced cameras, Evans says, so these cards and the people holding them are a relic of the past — one that she often finds hilarious, as some of the young people holding signs sported bell-bottoms and other bygone fashions, and sometimes, in hot weather, didn't wear much at all. Evans has taken to collaging images of extra clothes on them when she uses these old photos for public presentations. But in October 1980, this guy was warmly dressed, his completely unnecessary waders gaping away from his khakied hips.

I came to see Evans because this transformation from desert to wetland is taking place all over northeastern Nevada — and she's been involved since the beginning. Restoration ecologist Jim Laurie, a founder of Biodiversity for a Livable Climate, had told me that Evans and a group of bold thinker-doers — ranchers, other agency people, and scientists — have shown that it's possible to rehydrate the American West. 'For hundreds and hundreds of years, water has been draining off the continents,' he told me in a brief phone call. 'Humans keep wrecking everything that nature does to slow down water and keep it on the land, and the continents are desiccating.' Carol and this group, he said, were working to reverse that.

Rehydrate the American West? This piqued my interest immediately, since I grew up in California's Sacramento Valley and experienced what I considered its utterly unappealing dryness firsthand. I loved to run around barefoot as a child but had to be careful. The pavement in summer was so hot it burned my feet. The wild grasses dried into tiny lacerating spears; if I stepped into them, my mother would be at my feet with tweezers and a needle, its tip still hot and black from being held to a flame. I don't recall there being any of the wildfires so prevalent in the West now, but the fathers in our neighbourhood always took steps to forestall them; at some point after the grasses dried out, they'd conduct a controlled burn along the streets, armed with garden hoses and gunnysacks to remove the fuel from a flicked cigarette or other incendiary. We drove frequently from our house in Oroville to Lake Tahoe, and I was always eager for the landmark that told me we weren't far from the lake: a sign with a picture of Smokey Bear and a rating of the level of fire danger that day.

Rehydrate the West? It seemed to me that it had always been dry.

But recent memory — including my only somewhat recent memories of growing up in the 1950s — can be misleading. White settlers heading for California in the mid-1800s passed through the Elko area, and their written records indicate a rough landscape ribboned with lush creeks and wetlands. I grew up fascinated by the story of the ill-fated Donner Party, about half of whom died of starvation and exposure in the Sierra Nevada mountains in 1846. My family often drove over Donner Pass toward Lake Tahoe in the winter; as our car slid on the icy highway, despite my father's inexpertly applied snow chains, I feared our bones would wind up resting on top of theirs. The Donners and members of their wagon train passed through these lands that Carol Evans has been documenting, arriving after a scoundrel land speculator named Hastings directed them along a so-called shortcut that added weeks of bitter struggle to their trip. They staggered away from the Hastings Cutoff into a broad valley watered by creeks jumping with fish.

Where the Donner Party emerged from the cut-off, a rancher named Jon Griggs now manages the Maggie Creek Ranch for a non-resident family. Like the Maggie Creek Stream Survey Station S-9, the creeks on Griggs's land have transitioned from bare-banked trickles that dried up in some places in the middle of the summer to active streams well clothed with riparian vegetation and a more consistent, steady flow throughout the year.

Recent memory had Griggs fooled, too. He came as a

cowboy to the Maggie Creek Ranch in 1991, when the then ranch manager had already started working with Carol Evans on some of the changes that would lead to the transformations she's now documenting. Griggs didn't quite get the fuss about these creeks because he assumed that it was normal in these parts for them to be bare of vegetation and to dry up in the summer, to cut so deeply into the valley soil that they trickled along at the bottom of gullies. Then he read a memoir of Nevada ranching life in the early 1900s called *A Long Dust on the Desert*. The author recalls talking to an older rancher who arrived in the area in the 1860s who told him that the creeks once flowed on top of the land, not in gullies, and that they often overflowed and provided great pasture for the cattle.

'That's a powerful piece of intelligence,' Griggs told an audience at one of Jim Laurie's conferences. 'The creeks used to run on top of the ground. That told me why we should be doing this.'

The 'this' was a change in grazing that was both simple and tricky to get right.

Back in the 1970s, pretty much everyone agreed that some 100 years of cattle grazing had caused a host of environmental problems, including the degradation of these creeks. Some ranchers scoffed and said that the streams had always looked like this, but others felt quietly desperate that the work they loved might be ruining the land they loved, as they were often grazing cattle on land their parents and grandparents had settled and ranched. Riparian areas comprise a mere 1 per cent of Nevada's landscape, but neither the original homesteads nor modern ranches could survive without those threads of water. So all the ranches in this part of the country are built around slivers of privately owned land near the springs, creeks, and rivers, and then, through contracts with the BLM and the Forest Service, include nearby allotments of public land that expand the ranch to tens or hundreds of thousands of acres. Nevada's wide-open spaces are stark and seem like either a hostile moonscape or a welcome windswept break from the rest of the world, depending on your tastes. The people who choose to live there find it beautiful. 'It's empty because it's perfect,' rangeland ecologist Chris Jasmine told me when he and Jeff White, both employees of Nevada Gold Mines, took me to see a creek on one of ranches that the company owns atop its mines. 'We get nervous when there are too many trees and we can't see for thirty miles,' added White.

Despite their grief and fear for the future, ranchers continued to do as their ancestors had done, turning the cattle out onto the range in the spring and bringing them back when the snow came down in late fall — a roughly eight-month stretch in which the cattle were out of their control. The creeks and rangeland itself continued to deteriorate. Cattle can also degrade the uplands: they can overgraze their favourite grasses and ignore others, killing their favourites with their appetite and leaving behind bare ground that easily erodes. But in stark, arid Nevada, the cattle are always eager to hang out near the creeks, especially as the uplands brown off in the summer heat.

'They like to loaf around there because it's cooler,' says Tamzen Stringham, a rangeland and riparian ecologist from the University of Nevada, Reno, who studies changes in the creeks. A mind-blowing quirk of fate has brought her here: she is a descendent of Tamsen Donner, the wife of the Donner Party leader, who wrote to friends about her pleasure in 'botanizing' around creeks along the trail.

Cattle will happily graze the plants near the creeks to the ground, leaving the riparian area weakly vegetated or even bare. Their hooves can then inflict additional damage. When creek banks are bare and wet — and some stay wet all year — the cattle's hooves can sink as deeply as twelve inches into the soil. Many of the plants that stabilise riparian areas — grassy plants like sedges, bulrushes, and rushes — grow from rhizomes, a swollen piece of the stem just under the soil's surface. Without careful management of the cattle, their hooves can fracture the rhizomes and damage the roots. The overgrazed plants struggle to put out leaves to make more of the carbon-based fuel they create through photosynthesis, but they can't muster the resources to do so with their roots or rhizomes damaged. Through the double whammy of heavy grazing and hoof action, the vegetation around the creek dies. Without roots to hold the banks in place and plant biomass to slow down the water, periodic high water rips the banks away.

The damage to these riparian areas then radiates through the watershed. For water to seep into and nurture its surrounding landscape and to recharge groundwater, it has to move slowly. Sturdy vegetation-anchored banks that route water through twists and turns and spread it into occasional ponds enforce that essential slowness and seepage. As Evans explains it, they dissipate the water's energy. But when the vegetation dies and the banks degrade, the creek can lose its twists, turns, and ponds, and then the water rushes through the landscape without penetrating it. The fast-moving water erodes and lowers the streambed, incising deep channels into the earth. That causes the surrounding water table to drop and inhibits groundwater recharge; areas that used to be wet meadows become dry terraces with deeply incised creeks, the water so far below the land's surface that it can't support much of the life above. The creeks lose the veil of vegetation that hides water from the sun, the lowered water table prevents a cooling influx of groundwater to the creeks, and the creeks heat up to make a hostile environment for fish.

While Evans and I looked at her photos of the changing creeks of northeastern Nevada, she also pulled up satellite images of some of these dry terraces and pointed to the faint scars left by bygone creeks. We've all seen these scribbles across the land from the windows of a plane. 'These used to be flood plains — streams used to flow all over the valley — and now they're dry terraces,' she said. 'If you go back a couple of hundred years, many of these dry sagebrush terraces were a diverse mosaic of watered channels, beaver dam complexes, and riparian and wetland plant communities.'

Some environmental groups are eager to get rid of all

grazing on public lands, and I'm sure I would have felt that way myself before I started seeing how farmers, ranchers, and other careful stewards of the land can heal landscapes. It seems logical, after all: if the cattle are degrading these streams, then just get rid of the cattle and let the land begin to heal itself.

But in the 1980s, rangeland and riparian ecologists started to talk about a new approach to grazing that might help the creeks heal. Some were influenced by the ideas of Allan Savory, a biologist and former game-preserve manager from the former Rhodesia who had his own grazing epiphany in the 1960s. (I wrote about Savory in *The Soil Will Save Us*.) Savory had observed the African range become degraded from the action of cattle and other grazing animals and famously said he wanted to shoot every cow. But a period of careful observation and a study of the work of earlier scientists like André Voisin led him to the conviction that grazing didn't necessarily lead to degraded landscapes. Every landscape evolved with animal impact — and rangelands, in particular, evolved with the presence of grazing animals. Those ancient herds kept the land healthy by removing the build-up of dead grasses that blocked the sunlight from new vegetation and by disturbing the soil *just enough* — for water to penetrate, for seeds to find a footing, and for the cows themselves to fertilise the land with their microbe-rich dung. Savory argued that modern grazing degraded land because the cattle were turned loose and allowed to camp out wherever they pleased, for as long as they pleased, whereas ancient herds clumped together and kept moving in a bunch because they were fearful of predators. When humans domesticated them and removed that fear, their behaviour changed. He argued that it was that changed behaviour that was degrading the land.

Savory and his ideas are controversial, especially among environmentalists, but more and more ranchers and range ecologists are convinced that grazing done right can heal the land — and, paradoxically, that removing animal impact entirely can harm it. All landscapes evolved with and benefit from various forms of disturbance that stress the system — whether it is the occasional flood, periodic fire, or the action of big hungry animals — but these stresses can also promote new vigour, much as the disturbance of exercise promotes vigour and growth in muscle and bone.

'Herbivory has been around since the beginning of plants,' says Sherman Swanson, a rangeland and riparian scientist, also from the University of Nevada, Reno. 'In the Pleistocene, there were twenty-seven genera of megafauna, and many of them were large herbivores. In a place like Nevada, riparian areas were probably a centre of their life, just as they are for cows.'

Swanson points to research that shows that eliminating grazing from the land — the supposed panacea sought by some environmentalists — can actually make things harder for native fish. Scientists studied the impact on springs in Ash Meadows in the southwestern United States and Dalhousie in central Australia after they were fenced off from livestock.

Without grazing, the vegetation around the creeks burgeoned, diminishing the amount of open water and causing native fish to struggle. People tried to maintain open water by removing the vegetation by hand but couldn't keep up with the overgrowth. As a result, at least one fish population became extinct in Ash Meadows and eighteen fish populations — representing four out of the five native species there — went extinct in the Australian creeks. The researchers concluded that the creeks needed the 'substantial disturbance' of grazing to preserve habitats for native fish.

Savory's ideas fit with other observations about the positive impact well-managed grazing could have on the land — Swanson recalls that one of his professors used to say that if you wanted to find wildflowers, go to a meadow that has been grazed because the cattle have not only fertilised it, but also removed the thatch, allowing wildflowers to grow.

Then, in 1985, the BLM — led by riparian specialist Wayne Elmore from Prineville, Oregon — tried a new approach that incorporated some of Savory's ideas at Bear Creek in central Oregon, where cattle had been grazing since the late 1800s. The creek was terribly degraded and had dug a deep incision into the land, with eroding banks and only intermittent flow. Both the BLM and the Forest Service regulate the number of grazing animals allowed on their leased land and the period of time they are allowed to be there, and this particular piece of land had been licensed for twenty-five cows (or, using their terminology, seventy-five animal unit months or AUMs) from

June to August. Instead of allowing the animals to roam freely throughout the summer beginning in June — which is the traditional approach and one still practised on some ranches — they began allowing the rancher to let his cattle graze as soon as the winter snows melted, which saved him $10,000 on what he annually spent for winter hay. By early May, the rancher moved his cattle to another pasture to let the riparian and upland plants recover in time for the summer rainstorms. With this approach, the struggling riparian plants had plenty of time to regrow after a short but beneficial period of animal impact. As the creek banks stabilised, new plants sprouted from seeds that were either dormant in the soil or that floated in as part of the 'seed rain' — from birds dropping seeds, mammals shaking off the ones clinging to their fur, and willows that waft their seeds through the air on downy parachutes.

Over the next twenty years, this approach had a stunning impact on Bear Creek. As Elmore wrote in an article for *Range* magazine, by 1989 the BLM allowed the rancher licensing the land to run five times more cattle on the land than in 1976. By 1996, the riparian area had grown to twelve acres per mile of stream and was now producing approximately 2,000 pounds of forage per acre. The new growth of vegetation along the stream wasn't only providing more food for the cattle; it was also catching and holding the sediment moving through the creek, so much so that the stream bed was now two and a half feet higher than in 1976. The creek had also added the kinks and turns that slow water down and hold it in the landscape

— it was actually one-third of a mile longer, counting all those extra meanders — and researchers determined that the riparian areas were storing nearly four million gallons of water per mile by 1996, compared to 500,000 gallons in 1976. Rainbow trout were once again swimming in the creek.

Range scientists like Swanson began to evangelise sustainable grazing practices throughout the West, and agency personnel like Carol Evans — who came to the BLM as a fisheries biologist — developed close relationships with the ranchers willing to try making a change. Simple as it sounds, though, it was hard to break with the ranching traditions of their forefathers. It's much easier to let the cattle loose on a huge stretch of range in the spring and move them back in the fall than to actively manage their weekly or monthly whereabouts on the landscape, which this new approach required. Some ranchers tried to manage their cattle by herding them, but modern cowboys have health benefits and 401(k)s — they're expensive — and the stockmanship skills for deftly managing cattle are in short supply in the United States. Most ranchers and agency range managers turned to fencing that converted huge chunks of range into a series of smaller pastures, even though they hated the fences and their maintenance — they're expensive and cows seemed to regard them as a personal challenge.

Making this change also required ranchers to study the land in ways they had never done before, paying close attention to the growth cycle of the grasses and other plants. 'We were cattle ranchers, and we wanted a healthy cow,' says Agee Smith, who ranches 33,000 acres about fifty miles northeast of Griggs, up near the Jarbidge Mountains. His father went to the National Cowboy Poetry Gathering in Elko in 1994, heard Allan Savory speak, and came back to the ranch eager to try out his ideas on a ranch that was headed toward bankruptcy using the conventional methods of cattle rearing. 'Everything we did was around the cow. I went to college and got a degree in animal science, but I never learned anything about plants. As long as we had enough grass for the cattle, we never looked at the ground. It was ridiculous.'

A handful of ranchers in the region began to actively manage their cattle in more sustainable ways, and their creeks quickly began to show improvement. Grasses and rushes and willows sprouted on formerly bare banks. Griggs — still a sceptic — became a believer when he noticed that streams that used to dry up in midsummer began to maintain their flow all year. Then the willows grew into thickets and made his job managing cows even harder — during the times they'd graze near the creeks they'd get inside the thickets and it was tough to move them on. Still, he stuck with the program.

Then the beavers showed up around 2001, drawn by the abundance of willows.

Not that beavers had ever been completely absent from northeastern Nevada, but their numbers had diminished over the past two centuries. Fur trappers made a good living catching beavers in these parts as far back as the 1820s and nearly drove them to extinction — in fact, Peter Skene Ogden's 1828 expedition was instructed by the Hudson Bay Company to create a 'fur desert' and kill as many beavers as possible ahead of rival American fur traders. But even when the fur traders became a relic of the past, ranchers themselves were just as hard on the beavers, shooting them and blowing up their dams. They were worried about the beavers messing up ranch irrigation systems. This new cohort of beavers started gnawing down the stands of willows to build their dams, and Griggs was afraid they'd undo all the work that he and other ranchers had done to revegetate the creeks. And there was a bigger philosophical issue, too. 'We ranchers try to control the things we can control because there's so much we can't control,' Griggs told me with a sigh. 'It was hard to let go of that with the beaver.'

He watched in dismay as the beavers thinned out the willows and began building dams along Susie Creek. If it had been hard to get the cows out of the willows, it was that much harder to chase them out of the marshes and islands created by the beaver dam complexes.

But even though the willows themselves took a hit, Griggs, Evans, and the others saw that the beaver dams dramatically accelerated the recovery of the creeks and the surrounding land. Water that once shot through the landscape without wetting it was captured in ponds behind the dams. The water grew deeper and the zone of riparian vegetation steadily expanded into the surrounding landscape, adding not only more pasture and forage for cattle, but also a better habitat for wildlife. The water table rose. The iconic Nevada sagebrush that used to grow near the creeks started to die because they can't take a lot of moisture; sagebrush are adapted to drier soils. As the upland shrubs turned into bare and bleached branches, they were replaced by plants suited to a wetter environment. Without any help from the humans, the beavers had skilfully reengineered the landscape so that it now featured vibrant wetlands where there was once dust and gravel.

In addition to Evans's work measuring and recording changes on the ground, remote sensing has become a powerful tool for quantifying the ways that changes in livestock grazing, the resulting growth of riparian plants along the creeks, and the explosion of beaver activity have improved the watersheds. Open Range Consulting from Park City, Utah, used satellite and aerial images to show that riparian vegetation increased by more than 500 acres on a section of Maggie Creek between 1994 and 2016. In another survey of Susie Creek conducted by Trout Unlimited from Boise, Idaho, riparian vegetation increased by 100 acres, the amount of open water increased by twenty acres, the length of the creek increased by three miles (counting all those new twists and turns), and there were now 139 beaver dams when there had previously been none. The improvements in these watersheds continued, even when the weather tanked. In 2015 — four years into the worst drought any of the ranchers had ever seen — portions of creeks that used to dry up midsummer still flowed throughout the year and were surrounded by lush vegetation.

Now Griggs considers beavers the ranchers' friend. 'You can spend as much money as you can to replicate what they do, and you'll fail miserably,' he declares. 'Now we have this subterranean bowl of water here that they created.'

But the beaver couldn't have done this without the change in grazing, which set up the preconditions for their success. Agee Smith recalls that beavers used to build dams on Cottonwood Creek and other small waterways, but that the dams routinely failed during the spring runoff. The surge of water would hit a beaver dam and then wash around it, exploding the degraded banks of the creek and sending a rush of sediment downstream that would ultimately clog the ranch's irrigation system. The dams themselves would break up in the process. 'This scenario was happening forever,' he says.

Ten years ago, Smith noticed that the beaver dams had stopped washing out so often and that the banks around them didn't erode in the spring runoff. The growing health of the riparian areas around the creeks now allowed the beavers to link their constructions to solid stands of living vegetation. Even when there's a big surge of water, Smith and his family find that the water downstream remains clear because the sediment gets trapped behind the dams and settles to the creek bottom. 'We hardly have to clean out the filters to our irrigation system anymore,' he says. 'The beavers are great filters for the whole system.'

Humans are often eager to impose our own engineering

or technical solutions on a landscape, but with patience — and it does take patience — nature and the action of plants and animals can repair the damage. 'We don't always have to do the repair,' says Tamzen Stringham. 'Sometimes we are too aggressive. Sometimes our engineering does more damage than good. Mother Nature can do the repair.'

The story of the reborn riparian areas in northeastern Nevada is more than just the story of how humans can work with nature to heal landscapes, although that's a really great story. It's also the story of how humans who used to view each other with suspicion or even outright hostility have learned to work together.

This is not the story that most of us expect from Nevada, where the most nationally famous rancher is Cliven Bundy.

Hundreds of miles south of where Griggs, Smith, and other ranchers are working with people from federal agencies to restore riparian areas, Bundy and his family have been feuding with the federal government for decades. Bundy refused to pay the fees for his cattle to graze on a BLM allotment his father had been granted, and the family accrued close to $100,000 in unpaid fees by 1995. He compounded that defiance by allowing his cattle to wander BLM land that was completely closed to grazing. After two courts ruled against him, the BLM declared they were going to remove his cattle from federal lands in 2014. The roundup turned into an armed standoff between a militarised BLM and the Bundys and armed members of antigovernment groups that came to support them. When a

CNN crew asked Bundy whether he was going to pay his fees and fines to the federal government, he retorted, 'I have no business with 'em.' The government's case against the Bundys was dismissed in January 2018, and it lost an appeal of that decision in 2020. The Bundy cattle are still grazing on public lands without paying for the right to do so.

While none of the ranchers and agency people involved in the riparian repair effort had an extremely fractious relationship, it wasn't always a warm one, either. It became warmer as the creeks showed improvement and everyone took heart in this joint achievement, but the relationship really transformed when Agee Smith and his wife, Vicki, decided they wanted to turn their whole ranch over to Savory-style holistic grazing in the mid-1990s. In the process, they created a model for rancher–agency–scientist cooperation.

Agee Smith's great-grandfather Horace Agee bought the Cottonwood Ranch back in the 1920s from the O'Neils, a family who moved to California during the Gold Rush, and then, failing at gold mining, moved to Nevada and developed a reputation for cattle rustling and gunfights. It's a beautiful and remote high-elevation property in what's now called the O'Neil Basin, cupped by the jagged fingers of the Jarbidge Mountains to the northwest and the Snake Range to the east. I bumped over thirty miles of dirt road to get there and was talking on the phone to a friend in Ohio when three mule deer ran across the road in front of my car, followed a minute later by a mountain lion. ('Why did you shriek?' my friend asked.) Smith's mother lives in the log cabin built during the O'Neil years, Agee and Vicki live in another house, and their daughter, McKenzie, and her husband, Jason, live in a third. When Agee was growing up, his mother drove nearly a hundred miles each day to take the kids to school. Now there aren't enough local kids for there to be a school, so McKenzie and Jason homeschool theirs. 'Ranches now have fibre-optic Internet,' Agee tells me. 'We're in hog heaven. We're connected to the world.'

For decades, the family ran cattle on the ranch and hosted deer and elk hunters in the fall. In the 1980s, their cattle business bottomed out and they made a living from the hunters — building a magnificent lodge, with the help of new business partners — and ran other people's cattle on their land, with only a few of their own. But their creeks were looking worse and worse, and their federal landlords — the BLM and the Forest Service — were threatening to fence off the ranch creeks from the cattle. 'Financially that wouldn't have worked,' Agee says. 'Emotionally it was a tough time. We didn't even want to see the Forest Service or BLM people when they came around.'

That's when Agee's father heard that a fellow rancher was going to read some of his poems at the Elko National Cowboy Poetry Gathering in 1994 and, intrigued, he went, and came home telling Agee he should look into Savory's work.

'To hear that grazing and livestock can rehabilitate the land was life-changing,' Agee told me as we sat at the dining table in their hunting lodge, taking in the sight of one horse wandering through the Smiths' yard nibbling at shrubs and fallen apples, against a backdrop of dozens of other horses swishing around a corral, against a grander backdrop of the Jarbidge Mountains where their daughter was guiding a group of elk hunters. This part of the country seems to accordion away in one dramatic vista after another. 'That cattle could be used as a tool; that they weren't the bad guy all the time. I had bought into the idea that we were in a business that was harming the environment, and I didn't know any different or how to change things. But we took a class [run by a former associate of Allan Savory] and started going down that path. We will forever be on this journey.'

The Smiths offered to put their ranch up as an experiment in this new approach — it was the only thing they thought might save their cattle business, and besides, they were now excited about this whole new way of looking at nature. Amazingly, the state and federal agencies all agreed that it was worth a try. The Forest Service and the BLM put the ranch on a special status that relaxed some regulations and granted the Smiths greater flexibility in how they managed their herd. Further, the agencies agreed that the Smiths could expand the herd from 300 cows to 1,000 because this new approach turned the old thinking about livestock on its head: instead of being a scourge, it held that well-managed cattle had a positive impact on the land and that you needed enough of them to do the job right.

Key to the Savory approach is the pulling together of

stakeholders, so that every person or group affected by the project has input. In 1997, as the project began to take shape, the Smiths held a meeting of some fifty people on the deck of the hunting lodge. They invited everyone who had an interest in this piece of land, from neighbouring ranches to a wide range of agencies and environmental groups, even the ones that hated ranching and wanted to drive it from public lands. The first question posed to everyone on the deck was, 'What do you want this land to look like in ten years? Twenty years?'

Everyone wrote their answers on index cards, and then came the grand moment when the cards were unfolded and the answers read. One after another, the answer was the same: everyone wanted a healthy landscape.

This group became the Shoesole Resource Management Group — named for a historic cattle brand used in the area — and consolidated into a team that met monthly for the first few years (and now meets three times yearly). From the start, the meetings were managed in such a way as to break down the barriers between the participants, using methods developed by Bob Chadwick, a former Forest Service forest supervisor who learned about conflict — and the surprising opportunities for consensus — during the spotted-owl controversies of the 1990s. In his book *Finding New Ground: Beyond Conflict to Consensus*, he said, 'I found there was a simple equation: If people disrespect each other, they will disrespect the land.'

In each meeting, the Shoesole group sits in a circle of chairs, avoiding the visual perception that anyone is at the head of the table and thus the authority. Trained facilitators run the meetings to help navigate thorny, hot-button issues. Everyone gets the chance to talk without being interrupted. At meetings about public land management, there is often only the mildest of consensus among the participants in the beginning, and everyone naturally tends to think they are the aggrieved and misunderstood party. Here, they are encouraged to reveal the raw and tender feelings disguised by a clenched jaw or stiff upper lip.

'No one loves being vulnerable, but it shifts things,' says Laura Van Riper, a BLM social scientist, trained by Chadwick, who specialises in conflict resolution, collaboration, and consensus-based facilitation on behalf of the National Riparian Service Team. She has worked with many of the Shoesole participants as well as other groups. 'Every consensus-based workshop gets emotional at some point. Hell, I cry at all of them.'

This kind of vulnerability is difficult for the ranchers and cowboys, who suspect others don't appreciate the difficulty of making a living from the land or understand their love for it. Difficult for the bureaucrats, too, Van Riper explains, who are often as frustrated with red tape as the ranchers and are often stuck behind computers, away from the land that they, too, love. Difficult for the scientists, who don't understand why everyone doesn't respect the science. 'One professor of biology told me that wildlife management is 99 per cent about people and only 1 per cent about wildlife,' Van Riper says. 'But natural resource managers learn about that 1 per cent for four years, not the human part. I hear them say that we need to be logical and science-based about this work, but that's not how we're going to solve these problems.'

With the team's support and the help of the agency staff running interference with bureaucratic rigidity, the Smiths abandoned conventional ranching practices and embarked upon the very new experience of getting to know their land. 'I thought I knew this land,' Agee Smith says. 'I grew up here, but I didn't know what I didn't know until we started this process. You crawl around on your hands and knees to set up monitoring plots. You look at grasses, and how what's going on in the soil affects the grasses, and how that affects our animals and the wildlife, and how we can influence all that for the positive or negative.'

The Smiths began to manage the cattle with electric fences and cowboys (including their then ten-year-old daughter McKenzie), moving the cattle in and out of the riparian areas depending on the growth stage of the vegetation there. That meant nudging them along — gently, so as to not stress them — from one pasture to another every seven to fourteen days. The cattle had their own ideas about where they wanted to spend the season, and the Smiths quickly sold the ones that insisted upon hunkering down by the creeks all summer. They trained the others to move into pastures that needed more animal impact by dropping salt licks there. The first five years were tough, and fire — a frequent threat in these parts — devastated parts of the ranch twice. But the land sprang back more quickly than before. They noticed increasing numbers of wildlife — more sage grouse, more antelope, more elk. And they observed the same kinds of changes in the creeks that Griggs and the ranchers near Elko had seen: the arrival of willows, the proliferation of beaver dams, the water creeping into the rest of the landscape, the arrival of wetland plants, the retreat of the sagebrush and other dryland shrubs. The creeks that had been rated poor or fair by government scientists were all rated excellent or good by 2011.

I wasn't in northeastern Nevada when Shoesole was holding a meeting, but I did manage to attend the meeting of another group of ranchers, agency staff, and scientists that have been meeting since 2012. Inspired by Shoesole and including some of the same people, the SANE group — Stewardship Alliance of Northeast Elko — includes eight ranches; altogether, the Shoesole and SANE groups represent 1.7 million acres. SANE was launched by rancher Robin Boies and Connie Lee, a biologist with the Nevada Department of Wildlife, to see if the group could come up with a plan to protect sage grouse before the federal government declares it 'endangered', a step that would usher in a new raft of regulations further restricting ranchers' ability to manage their land and animals.

When I arrived, the participants were just settling into a big clunky square — they did the best they could to circle up some rectangular tables — and then immediately stood for a serpentine greeting. One by one, they peeled away from the table to speak to the person next to them and moved in a line along the outside of the circle, then did it again going the other direction. So, everyone had two chances to greet each other before we settled down again. Then I sat, mute and dumb as a brick, while they talked about various projects. I tried to figure out who was who, but with one exception — a lean cowboy with a stunning scarf held in place by an equally stunning scarf slide — I couldn't tell who was a rancher and who was an agency employee or scientist. It was hard to believe this group hadn't always had the easy intimacy and warmth I witnessed that day. I learned later that this level of trust had only developed because people like Boies and Lee had pursued it. They believed it mattered more than anything else, and they persisted in creating the conditions for it to thrive.

# Chapter 5

# Agriculture That Nurtures Nature

The entomologist Jonathan Lundgren and I stand at the edge of Roger's cornfield on a cool August morning in the Prairie Pothole Region of South Dakota. We're supposed to meet a group of his students, but there is no sign of them. Perhaps they are on their way, I think, then Lundgren cups his hands around his mouth and sings out, 'Marco!'

No one answers, but he nonetheless springs off the gravelled edge of the road and strides into the field next to the corn, carpeted with a tangle of low-growing greens and golds, which I try not to crush as I follow. 'Marco!' he calls again, and then again. Finally a voice responds from deep within the muffled confines of the cornfield itself. 'Polo!'

Not that it's all corn in this cornfield. Roger is one of the innovative farmers who has invited Lundgren and his team onto his land to conduct research into the benefits of regenerative agriculture, generally defined as agriculture that builds soil health and overall biodiversity and yields a nutritious farm product profitably. Where most of America's nearly ninety-two million acres of corn feature armies of a single genus arrayed in military precision with identical thirty-inch corridors of bare, baked dirt between them — these single-genus plantings are called monocultures — it's hard to see Roger's corridors. Each hosts its own jungle of plants — buckwheat, field peas, hairy vetch, lentils, flax, millet, sorghum, sorghum-sudangrass, and cereal rye — and as we approach the field, some poke through the outer edge of corn, like rowdy vegetative inmates straining against cornstalk bars. High above the corn tassels, sunflowers loll their yellow heads.

It takes several more iterations of Marco Polo before we narrow in on the location of the students. Even after Lundgren and I push our way in, there is so much vegetation that we can't see them until we're only a few feet away. There, they sit between rows of corn, long clear tubes dangling from their mouths. They've spent the morning sucking up insects from soil transects and collecting them in a container so they can tally up the number of beneficial bugs all those non-corn plants have drawn to the field. While conventional American agriculture — and even many large-scale organic enterprises — follow the monoculture model, Roger is trying out the robust biodiversity of a polyculture in this field.

Jonathan Lundgren is an independent scientist whose partners and funders are farmers, but until a few years ago, he was an entomologist with the US Department of Agriculture (USDA). He was hired right out of grad school in 2004, at a time when farmers were reeling from the impact of a soybean aphid that found their fields an inviting target — as nearly all of them were planting huge soybean monocultures, it's almost as if they were inviting the pest to dine without the distraction of any plants not on their preferred menu. Lundgren began pursuing two related fields of study: whether the aphid could be defeated by diversified systems — meaning, fields in which farmers encouraged biodiversity over monocultures by adding other plants and even animals to their fields — and whether the pesticides most farmers used to control the aphids and other pests posed unnecessary risks.

Some farmers had tried bringing in biocontrols — packages of pest-eating insects — just as backyard gardeners send away for praying mantises or ladybugs. But the effort was largely a failure, Lundgren says, because there was nothing in these monocultures for the predator insects to eat other than the pest itself. The beneficial predators would happily eat pollen and nectar until the pest arrived, but farmers had been convinced — by their ag-school professors and professional agronomists as well as a farming culture that values neat-as-a-pin landscapes — to keep the corridors bare of any plant that might steal water or nutrients from the crop. And, of course, the purveyors of herbicides and the seeds for plants bioengineered to resist herbicides — usually the same powerful monopolies — drum a constant message of the necessity of chemical warfare against weeds. It's key to their business plan.

Lundgren's research showed that the chemicals farmers were using to combat the aphid — called neonicotinoids

— didn't increase their yields, as the big companies selling neonicotinoid-coated seeds loudly claimed. Rather, the neonicotinoids hurt the farmers. 'They were killing the natural enemies of the soybean aphid,' he tells me back at the Blue Dasher Farm, which he started as a living model farm with his wife and children to show that one can make a good living from regenerative agriculture — he is now a small farmer as well as a scientist. 'And the neonicotinoids don't even kill the soybean aphids, because they don't arrive in the field until the chemical is largely out of the plant.'

After he'd been working on soybean pests for a few years, Lundgren got word from his bosses at the USDA that they wanted him to shift his research focus to corn. 'I swore I'd never work in corn!' he groans. 'Corn is a miserable crop to work in. You get soaking wet in the morning, and when it's tasselling you get covered in pollen, and you itch like crazy.' But he knew it was an important field of study, as corn agriculture has such a huge footprint in the United States. This is America's largest crop, covering an area the size of sixty-nine million football fields, and the way farmers grow it has an oversized impact on our landscapes, air, water, and health.

Lundgren quickly applied himself to researching the corn rootworm, which American farmers spend millions of dollars on chemical pesticides to combat. He soon ran into a brick wall of dogma. His interest, as always, was not in looking for the right chemical to kill it, but rather trying to figure out how to help its natural predators eliminate it. Everyone told him the corn rootworm had no predators — even entomologists who had spent their entire lives studying this pest. Regardless, Lundgren was convinced that there had to be natural predators and that the other scientists were somehow missing them. He and his lab began trapping insects inside little cups stationed in cornfield soil, then smashing the insects open back in the lab and using DNA analysis to examine their stomach contents. The result, Lundgren says, was that 'we found rootworm DNA everywhere we looked. Everything was eating it! Ants, beetles, spiders, dozens and dozens of species.'

He found that farmers' problems with rootworm depended on the number and diversity of these predators. The rootworm has several defences, including a sticky toxin that gums up the mouths of chewing predators like beetles. They avoid the rootworm unless there's nothing left to eat. For this reason, Lundgren calls rootworms the orange creams of the insect prey world. 'You get a box of chocolates, and the orange creams are always the last damn chocolates in that box,' he says. 'You usually bite one and put it back. Same thing with the rootworm: only when the predators are abundant and all the other prey are eaten, that's when they'll finally eat the rootworm — the orange cream.'

Lundgren began meeting farmers who were attracting robust predator communities to their farms by planting cover crops — plants not intended for market, but which were traditionally used to protect soils from erosion and are now also being used to add biodiversity — on fallow fields and between the rows of their market crops. They weren't using pesticides and weren't having pest problems. They were even having fewer weed problems, as all that extra vegetation was also attracting beneficial insects that eat tiny weed seeds. He felt that he was witnessing the birth of a new approach to agriculture, and he wanted to support it.

At the same time, the problems with conventional agriculture were becoming more and more apparent to him, as he met commercial beekeepers whose bees were dying by the millions as they were trucked around the country to pollinate crops from one heavily sprayed field to another. And even though Lundgren's early years at the USDA were career gold — he ran his own lab, published a book and nearly 100 scientific papers, and won awards from both the agency and President Obama — he started feeling a chill within the agency after his paper was published in 2012 showing that neonicotinoids don't increase soybean yields. He was reprimanded over and over for minor issues, especially for talking to the press and the public about his research, then suspended twice. The hostile scrutiny mushroomed when he wrote a paper in 2015 showing that a neonicotinoid pesticide harms monarch butterflies. He filed a scientific integrity report with the agency, alleging that his work and communication with the media were being disrupted, and later, a whistleblower suit arguing that he was being unfairly disciplined to suppress his science.

'It became clear that the USDA wasn't interested in alternatives and that their job — and mine — was to protect agriculture as we know it,' he says. 'But the farmers were doing something different in spite of the USDA. I'd rather work with them.'

Lundgren left the government in 2016 to start the fifty-acre Blue Dasher Farm — named for his favourite dragonfly — as well as Ecdysis Foundation, a non-profit research lab located on the farm and named for the stage of metamorphosis when insects shed their skin. His connection to some of the country's most innovative farmers had blossomed into a community by then. He decided to embark upon agricultural research in a way seldom done — by having farmers fully participate in the science, fund it, and sometimes initiate it.

I visited him a year later, arriving in a gas-guzzling, cheese-yellow Dodge Charger, which the rental agency gave me when I said I wanted a small car and they said, 'Two door?' and I cluelessly nodded. I roared up the freeway, then roared onto the grounds of Blue Dasher — the car didn't do anything but roar — scattering chickens and cats and prompting one of his students to later complain, 'I thought you said she was an environmentalist!' By then, Lundgren was connected to a worldwide network of farmers. Sometimes he reaches out with a research question and the farmers offer themselves as participants in a new study. Sometimes a farmer or two will come to him with a question and he reaches out to the greater community to see if others want to join a study. While the lab work takes place in the Ecdysis building — an old milking shed that Lundgren transformed into a well-endowed laboratory with discounted shelving and cupboards from home-supply stores and donated beakers and other equipment — all the field work takes place on the farmers' land, not designated research plots that have never been part of putting food on the table. He usually has over a dozen different research projects underway.

Much of the rest of the country has signed on to the dubious idea that farmers — especially in places like South Dakota — are not interested in science and sullenly resist enlightenment. It's probably no truer than any of the other mean-spirited stereotypes flying around the country, and it's certainly not the case for the worldwide cohort of farmers who are trying to raise food and steward their land ecologically. That worldwide cohort is a larger group than the ones who now call themselves regenerative farmers, who are largely the scarred, savvy survivors of industrial agriculture in the United States and other so-called developed countries and who began to carve a different path to save their farms and their health. The larger cohort also includes Indigenous farmers around the world who have been respectfully cultivating their land and seeds for centuries; their kind of farming is often called agroecology. It includes Asian farmers who follow in the footsteps of Masanobu Fukuoka, the scientist and rice farmer who wrote the 1975 classic *The One-Straw Revolution*, relying on the fertility of the local environment and eschewing tilling and costly inputs; theirs is called natural farming. It includes the best organic farmers who are paying close attention to the health of their plants and soil.

All these farmers are citizen-scientists. They walk their lands with the informed, fond curiosity of naturalists and know that it's folly to approach their work as if they were baking the same cake every season using the same recipe and ingredients. They know that nature has many moving, changing, interacting, living parts and that these parts need our respect. For the farmers trying to find a path to both healthy profits and healthy landscapes, Lundgren's science can answer some of their questions about how to proceed.

'The farmers I work with are really interested in science,' Lundgren says. 'For the farmers who want the status quo, an infrastructure of science and education is all lined up for them. But there is no infrastructure for the ones who are trying to innovate, and that's who I'm tailoring our work to: the farmers who are really trying to push it.'

Status-quo agriculture has a ruinous impact on the 4.5 billion acres around the world devoted to row crops and the many billions of acres of rangeland and pastures, as well as on the landscapes that surround them. For every pound of corn harvested in Iowa, for instance, more than a pound of topsoil is lost; for every pound of soybeans, two to three pounds. That lost soil drifts into our watersheds with the wind and the rain, carrying the chemicals conventional farmers have so diligently applied to their fields. Fertilisers poison our waters by altering the balance of microbial life there: nitrogen pouring into the Gulf of Mexico prompts the growth of algae, which sucks oxygen from the water, causing a massive dead zone for fish and other living things; phosphorus pouring into Lake Erie prompts the growth of cyanobacteria, which squirt toxins into the drinking water. The total annual cost of erosion from agriculture in the United States is about $44 billion per year. On a global scale, the annual loss of seventy-five billion tons of topsoil costs the world about $400 billion per year.

We can rebuild that precious topsoil — molecular microbiologist David Johnson's research suggests that ten tons of soil carbon per hectare can be rebuilt annually — but only by mimicking nature to restore the soil's microbial diversity. Through photosynthesis, plants pull carbon dioxide out of the air to create the sugary carbon-rich compound that fuels their growth, but they don't keep all of it. Instead, they strategically leak some 40 per cent of that fuel through their roots to feed their partner microorganisms in the soil. Johnson has found that plants can even direct up to 90 per cent to their soil partners when that community needs an extra boost — in other words, the plants disperse more of that hard-won carbon fuel to improve their shared environment. The microorganisms pay for that carbon bounty with minerals, water, chemical defences against disease and insects, and other things plants need. The microorganisms exhale carbon dioxide, just as other carbon eaters like humans do. But when soil is healthy and covered with vegetation, it stores some of that leaked carbon, often for very long periods.

Johnson has found that as soil becomes richer in carbon, its microbial community has the energy to take on bigger and more varied tasks. 'They're like us,' he says of soil microbes, 'they depend on energy to function. Without energy, they're basically cavemen.'

When soil is low in both fresh carbon from plant roots and stored carbon, the microbial population struggles to survive. But as carbon and the population of microorganisms increases, what was once basically rock dust — that's earth without the precious ingredient of life — becomes topsoil. As the population of microbes continues to grow beyond what Johnson calls their caveman stage, mutualisms and specialisations develop. Specialist microbes in the soil can do many things, including gobble up just about any pollutant thrown at them. 'All these chemicals that are building up in our environment can be a food source for them,' Johnson says. 'As long as they have a carbon-carbon, carbon-hydrogen, or carbon-oxygen bond, there's energy there for them.'

At this point, Johnson says, the bacterial colony can muster the kind of cooperative efforts that exist in human cities — the microbial version of traffic laws, hospitals, and community gardens. This large-scale cooperation enables them to use carbon more efficiently and can turn a piece of degraded land from one that's gassing off copious amounts of respired carbon dioxide into healthy land that actually retains carbon.

'We have to slow the respiration of carbon through the system,' Johnson says. In one of his field studies, he managed to increase soil carbon fourteen-fold, while respiration from the soil microbes only doubled. And this kind of super soil doesn't come at the cost of agricultural production; rather, it promotes it. On one of Johnson's plots — carved out of the New Mexico desert, inoculated with his custom-made, fungal-rich compost, and then managed in cover crops without tillage — sunflowers that usually grow five feet tall with six-inch heads grew seven feet tall with twelve-inch heads.

'Same seeds,' he told me. 'We don't even know what plants can do when they have the right microbes.'

Because of business-as-usual agriculture, lands that could be part of the climate solution are instead part of the climate problem. Soil that's disturbed and debased by status-quo agriculture and other human activities releases stored carbon — according to a 2018 report in the *Proceedings of the National Academy of Sciences*, agricultural land has lost 116 billion tons of carbon, with the most dramatic losses occurring in the last 200 years. Even as degraded as the world's soils are, though, they still store some 3,000 billion metric tons of carbon, according to soil biochemist Asmeret Asefaw Berhe. Every year, 4.7 billion metric tons of carbon dioxide — around half of what human activity releases into the atmosphere every year — are absorbed by the soil, the plants, and the seas, and most of it sticks in the soil. But imagine what healthy resilient soil could be doing for the planet! Continuing the kind of agriculture that ruins soil is a huge lost opportunity. Converting it to regenerative agriculture is the only way we will effectively deal with climate change.

Many regenerative farmers and ranchers have been growing

commodity crops and control hundreds if not thousands of acres of land, thus they have an oversized impact on landscapes. Their successes might help other farmers around the world who are currently being courted or coopted by industrial agriculture. It might help smaller practitioners resist industrial practices, and that would be a great thing — according to some estimates, 70 per cent of the developing world's food is actually produced by small farmers, not by megafarms. Regenerative agrarians have many champions and helpers, and Jonathan Lundgren is not the only scientist working directly with and for them. But his work is especially important because it is among the largest and most focused bodies of research on the differing impacts of conventional versus regenerative agriculture. Despite the many case studies and stories of success from the ranks of regenerative farmers and ranchers, professional doubters always say, 'Where's the science?' Lundgren's work can amplify the successes and help these innovative agrarians convince the rest of society to throw its weight behind this movement, instead of allowing a government under the sway of the big industrial agriculture monopolies to hinder it with status-quo policies.

Buz Kloot is another champion and helper. He's an aquatic scientist at the University of South Carolina who used to hate his work because he felt like a coroner: the waterways were dying and, he says, it seemed like the only thing he could do about it was to declare the cause of death. He didn't think anything could change, because he didn't think farming could change; he certainly didn't think the health of the soil in modern America's farmlands could change. Then he visited a farm owned by a soil-health pioneer, Ray Styer, who hadn't used chemical fertilisers in twenty-five years.

'Everything I knew said you had to put on fertiliser to make crops grow, but he told me he was just using cover crops to feed his plants,' Kloot tells me. 'I could understand every word, but I couldn't understand the concept. I couldn't understand how the biology alone could do this.'

As Kloot started to meet other farmers who were interested in pivoting away from the industrial model, he heard a lot of confusion about how they could wean themselves from chemicals without a precipitous drop in production. He found that university extension services' advice for fertiliser use varied wildly from state to state and seemed to be based on studies conducted decades earlier, when nearly all farmers were tilling their fields and weren't using cover crops. Farmers might spend $230,000 more on fertiliser depending on which side of a state's border they were on. But the regenerative farmers were creating new conditions on their land, and Kloot realised that they might have completely different fertiliser needs.

'If you have a chemical view of soil, you just think of soil as a place to grow plants,' Kloot says. 'With that view, the only way to make things grow is to manipulate them chemically. But if you see soil as a living, mutualistic dynamic ecosystem that's changing all the time, that's full of microbes, it's a whole different view and it affects the way you do things.'

Kloot went on to do his own experiments, with some

funding from his university foundation as well as crowdfunded cash from dozens of farmers and other small donors. For one study, entitled 'How Much Fertilizer Do We Really Need?' Kloot worked with regenerative farmer Carl Coleman to set up plots on Coleman's cover-cropped fields, some with no fertiliser at all and others with varying amounts. The results — which Kloot confirmed in a set of replicated studies — showed that regenerative farmers did not need as much fertiliser as was often recommended. Plus, they were steadily building up more soil organic matter — microbial, plant, and animal matter in various stages of decomposition. That organic matter contains both organic carbon and nitrogen, which soil microorganisms convert into a plant-available form near the roots. That suggested that the farmers' needs for supplemental nitrogen would continue to decrease in the future. And while conventional thinking predicted that the levels of potassium — another nutrient farmers typically add to their soils — would drop so low that the farmers might lose productivity, soil tests found that potassium levels fluctuated according to crop stage and weather.

'The farmers I work with haven't applied potassium in six years,' he says, 'but the prognosticators of crop failure continue to make dire predictions as these guys continue to yield the same or better and save a bundle on fertiliser. In our conventional, reductionist science, we've overlooked the ability of soil to supply these nutrients if we work with Mother Nature, not against her. The rules literally change when you allow soils to be living ecosystems.'

Regenerative agriculture encompasses a set of principles first articulated by a group of conservationists and educators from the USDA's Natural Resources Conservation Service — Ray Archuleta, Jay Fuhrer, Barry Fisher, and Jon Stika — that respect and support nature and that guide farmers to work with the many complex relationships that comprise nature instead of disrupting them. Nature is inherently complex, but humans often try to make it simpler for the production of food. Of course, farming can't help but disrupt nature — just about every human activity disrupts nature — but these innovative farmers are figuring out how to minimise the disruption and restore some of the natural diversity and complexity to their landscapes.

The principles for turning farmlands into living ecosystems basically subvert many of the practices used to simplify landscapes. First, farmers need to minimise disturbance, including biological disturbance, the physical disturbance of ploughing or tillage, and the chemical disturbance of fertilisers and the many '-cides' — herbicides, pesticides, fungicides, nematicides, and so on.

The grossest form of biological disturbance is the monoculture. Farmers usually try to compensate for the biodiversity desert caused by monocultures by rotating crops — instead of growing acres of corn in the same fields every year, they alternate the corn with soybeans one year and perhaps another crop the next. Crop rotation was a practice encouraged in the early 1900s by African American scientist George Washington Carver, who wanted to help poor farmers overcome the debilitating effect of cotton plantation monocultures. Until the Nixon administration in the early 1970s, when the government began pushing farmers into industrial-style production, farmers often rotated many crops through their fields. Carver also urged the planting of cover crops, including the nitrogen-boosting peanut. Regenerative farmers now routinely plant cover crops on fallow fields or between the rows of the cash crop and even let non-bothersome weeds play their part in restoring and nurturing biodiversity above ground and below.

The physical disturbance of tillage damages the communities of mycorrhizal fungi that support and connect plant communities. While agricultural land doesn't likely have the density of fungal life that ecologist Suzanne Simard found in forests — 300 miles of fungal strands under every footstep — tillage disrupts the fungal communities that are there and tilts the balance of the soil community toward bacteria. Bacteria are crucial partners in ecosystem health, of course, but soil dominated by fungi is even more vibrant and productive. 'Fungi have this unique trait of providing both logistics and communication,' says molecular microbiologist David Johnson. 'They create a communication network as well as one that can funnel or ship goods, including all the elements that a plant needs.'

Then there is chemical disturbance. Those of us who aren't

farmers have an instinctive aversion to the -cides, probably because we are leery of those chemicals clinging to the foods we eat. And some certainly do, as one of the many studies of the ubiquitous herbicide glyphosate (Roundup) showed that more than 90 per cent of pregnant women in a small sample from Indiana had glyphosate in their urine and that higher concentrations were associated with earlier deliveries in pregnancies. Most studies tend to investigate these agricultural chemicals' impact on human health, rather than landscape health, but there is evidence that they can disrupt natural processes in everything from the microorganisms in the soil to the microbiota of much larger creatures. For instance, researchers found that when exposed to nitrogen fertiliser for a number of years, the bacteria called rhizobia that 'fix' nitrogen undergo genetic changes that make them less beneficial for host plants. This not only affects the agricultural field where the nitrogen fertiliser is used, but also nearby areas that are subject to runoff. When agricultural lands erode and these chemicals leach through the watershed, they're disrupting this basic and essential plant–microbe mutualism wherever they go.

'Worldwide, the nitrogen cycle is off. We've changed it fundamentally,' says University of Illinois plant biology professor Katy Heath, who led the study with Jennifer Lau of Michigan State University.

The chemical disturbance to our landscapes disrupts other mutualisms, too. When I was visiting Lundgren, he suggested I talk to his beekeeper friend. I had the idea that this was just some guy with a few backyard hives. But after I roared into Bret Adee's driveway — he laughed when he saw my cheese-mobile and shouted, 'This is every nineteen-year-old boy's dream!' — and I saw the custom stained-glass windows of honeybees framing his front door, I had an inkling that he might have a few more than that. Turned out he was the president of the Pollinators Stewardship Council until 2020 and the biggest beekeeper in America, with more than 90,000 hives that he dispatches around the country to pollinate one crop after another: first almonds, then broccoli, avocados, cherries, apples, and more.

Adee's grandfather started the family bee business during the Depression, and the family was in it for the honey profits until the 1980s. Competition from Chinese and Latin American imports diminished those profits, and Adee shifted the focus of the business to crop pollination. He lost bees even back then — they range up to seven miles from the hive every day as they forage, so a beekeeper can't control all the dangers they might encounter — but he lost only 5–10 per cent of his hives every year, which was in the historic range. Beekeepers in non-agricultural counties still have losses in that range.

But in 2007 — a year after another beekeeper lost 90 per cent of his bees and coined the term 'colony collapse disorder' — Adee lost 44 per cent. His bees continued dying the next year. There were other ominous developments: queen bees were living six months instead of two to three years, and the fall supersedure — in which an orange-sized knot of bees departs a robust hive with a queen to form a new colony — stopped happening. As Adee frantically conducted research into what might be causing the problems, he kept running into mentions of neonicotinoids and Lundgren's work. Then he realised Lundgren lived nearby, and they became fast friends and collaborators, both chasing a solution to bee death and decline.

Adee agrees with Lundgren that landscape simplification — and, especially, the chemicals that are used to create and maintain these simpler landscapes — is the culprit behind the bee carnage. According to his reading of the scientific litera-ture, fungicides kill naturally occurring microbes on flowers and plants. Those microbes would ordinarily be introduced to the hives by returning bees and help them digest pollen. 'A lot of pollens have really hard shells,' Adee explains. 'Bees can't break down the shells without these microbes, so you can have a whole hive full of honey and pollen but the bees will starve because they can't access the nutrients.'

When bees encounter fungicides along with neonicotinoids, Adee says the synergies between these two chemicals amplify the combined toxicity so that they become seventy times more toxic than DDT. But bees are exposed to far more than just these two chemicals: bee scientists at Pennsylvania State University have found residue from over 100 pesticides in hives, and these disrupt the bee immune systems in different ways.

The weedkiller glyphosate is doubly hard on honeybees (and bumblebees). The simplification of the landscape, often achieved through herbicides sprayed widely among crops bioengineered to resist these chemicals and along rural roadways, means there's often not enough for bees to eat, making them forage more widely and work harder. And glyphosate sabotages them from inside, too. Monsanto (now part of Bayer), the company that makes this herbicide, has always claimed that animals from bees to humans aren't endangered by glyphosate, as it targets an enzyme found only in plants and microorganisms. But honeybees, like humans and most other living things, have a mutualistic relationship with microbes that keeps them healthy. Evolutionary biologist Nancy Moran and her colleagues have found that glyphosate diminishes the presence of eight big players in the bee microbiota and that bees with impaired microbiota are more likely to die when threatened by a pathogen.

That's why Adee is sceptical of one of the common hypotheses floated by government officials, industry, and some ag-school academics about the cause of bee death: the Varroa mite. This parasite lives on adult honeybees as well as on their brood, sucking away until the adults are weakened, causing malformation in emerging bees, and spreading viruses throughout the hive. But this pest has been around since at least the 1980s, and Adee says it never used to be the problem that it is today. Fifteen years ago, he recalls, he would test for mites by putting a cup of bees in a jar, adding a shot of ether, and shaking. The mites would fall off the bees and stick to the wall of the jar. Back then, he knew he needed to treat the hives if twenty to twenty-five mites stuck to the wall.

'Now the bees are so weak that if you see more than five mites, you know a train wreck is in the works,' he says. 'The bees have lost a lot of immune function, so that a few mites carrying viruses that have been in the environment forever become problematic. Bees are like us: if there's a cold going around, it might not be a problem for most people but it can be a death sentence to someone with AIDS.'

All this has turned Adee and his wife, Connie, into regenerative farmers themselves. First they stopped tilling their land, then they adopted the movement's second principle: they keep their soil covered at all times with what North Dakota conservationist Jay Fuhrer calls a magic carpet of either living plants or plant residue. The point is to mimic natural areas, where bare soil is rare. The living roots and decomposing plant residue feed the soil microorganisms as well as protect soil from summer heat, which can be thirty degrees cooler underneath the cover. That difference in temperature is crucial for the survival of both soil microorganisms and beneficial insects.

'Imagine shrinking to the size of a small beetle,' suggests Mike Bredeson, one of the graduate students who was sucking up insects in the middle of Roger's cornfield — he was the Polo to Lundgren's Marco — and has since finished his PhD. 'Without plant cover, you might be able to use that space during the night but not during the day. When that cover isn't there, these critters can't lay eggs, can't survive, can't perform the services we want.' And though some farmers worry that planting a cover crop will rob their fields of the moisture that their cash crops need, Bredeson says that cover crops wind up saving moisture. Bare soil allows massive evaporation, which draws up salts in the soil — often these are artifacts of agricultural chemicals — and reduces soil productivity. Cover crops' shade not only reduces evaporation, but the density of cover-crop roots creates millions of channels for water to penetrate deep into the soil.

The third principle of regeneration: farmers need to add as much diversity to their fields as possible — in other words, the greater the number of species in their cover crops, the better. Just as doctors advise us to eat from a wide variety of foods, having a wide variety of plants in a field offers a rich, diverse diet to the insects and soil microorganisms that rely on them. Regenerative farmers often maximise this diversity by making sure their cover crops have varied root depths as well as various heights.

The ultimate diversity — and the final principle of regeneration — comes from adding animals. All farms used to have animals, both for what they contribute in work and what they contribute in meat, eggs, and dairy, but farmers were prodded decades ago to adopt a factory model that encouraged them to concentrate on one or two commodities and ditch their other food-making enterprises. Of course, many landscapes have become degraded because of poorly managed animals (see chapter four about northeastern Nevada, for example). But regenerative grazing has the opposite impact. Farmers and ranchers move animals from field to field so that they have a beneficial impact on the land, with their microbe-rich dung and nitrogen-rich urine fertilising it, and their hooves pocking the soil surface just enough to maximise water penetration and push seeds into the soil. They leave the animals there long enough and in dense-enough numbers so they munch all the plants but not so much that they stop photosynthesis. This gives tired lands a biological jolt that leads to greater overall biodiversity.

That jolt can be profound, and research published in the journal *Nature* suggests it can last thousands of years. Scientists have recently concluded that wild African landscapes like the Mara-Serengeti were dramatically shaped by the work of ancient herders and their cattle, sheep, and goats. They demonstrate that wildlife hotspots in today's southern Kenya were created by the accumulated nutrients from the Neolithic nomads' grazing animals, which roamed the landscape by day and were kept in pens at night. Those formerly penned areas are now rich grasslands frequented by gazelle, wildebeest, zebra, and warthog, dense with the worms, dung beetles, and other insects that attract birds and reptiles. At least one species of gecko only thrives in these rich glades created by the penned animals.

Lundgren's research focuses on farms and ranches that are on the regenerative path — some embrace all these principles, others are making progress on some — and compares them with nearby farms and ranches practising conventional agriculture. When I visited, he was in the process of publishing a study conducted with his graduate student Claire LaCanne, now an extension educator in agricultural production systems at the University of Minnesota. The study followed ten cornfields per farm on twenty farms over two growing seasons. Half the farms were regenerative — they used a diversity of cover crops, never tilled or applied chemical insecticides, and grazed animals on their crop residue — and half were conventional. Of the latter, eight practised tillage, and they all used seeds genetically modified to resist insects and/or used seeds treated with neonicotinoids, while leaving their soil bare after they harvested their cash crop. The study tracked soil carbon, insect pests, corn yield, and profits.

The results, published in 2018 in *PeerJ*, a journal of life and environmental research, gave the imprimatur of science to the kind of success many regenerative farmers have reported for years. Lundgren and LaCanne found that there were more pests in the cornfields that were treated with insecticides and/or used GMO seeds than in the pesticide-free fields, presumably because the cover crops attracted battalions of prey insects that decimated crop pests — and because insecticides weren't killing off those beneficials. And while the regenerative farms used older, lower-yielding corn varieties with less fertiliser and had lower yields, their overall profits were 78 per cent higher than the conventional farmers'. Partly, this was because the regenerative farmers' costs were so much lower, with no cash outlays for costly insecticides and GMO seeds. They also 'stacked enterprises' and had two or more sources of income on the same acre — in this case, they grazed their cattle on corn residue after harvest and got a premium price for pastured beef. What was the primary factor correlating with farm profitability? The amount of carbon and organic matter in the farmers' fields, not their yields.

'The study really showed the benefit of thinking of your farm as an agricultural ecosystem,' Claire LaCanne told me.

Other studies are yielding similar results. In California, regenerative and conventional almond growers joined a more recent study conducted by another of Lundgren's students, Tommy Fenster. Before his connection to Lundgren and Ecdysis, Fenster was working for the Alameda County Waste Management Authority, teaching people how to make and use compost and learning about regenerative practices on urban farms. When he started a master's in biology at California State University, East Bay, he reached out to ask Lundgren to be his offsite adviser.

Fenster's is also a two-year study, following four regenerative orchards and four conventional orchards for one year, then tracking different sets of orchards over a second year. It was hard to track down regenerative orchards — the new certification for Regenerative Organic had not yet been extended to orchards — so Fenster cold-called growers around the state. He queried them about a list of regenerative practices for orchards that included avoiding synthetic fertilisers or pesticides; using compost, diverse cover crops, grazing animals, and hedgerows to attract and harbor biodiversity; and keeping the soil covered most of the year. Any orchardist who followed more than four of these practices was considered regenerative. All of the orchards that qualified were certified organic, although that wasn't a requirement for the study. Fenster then set about examining all of the orchards for insect crop damage, quantity of soil carbon and nitrogen, effectiveness of water infiltration (how quickly water penetrates the soil), composition of the invertebrate population (insects, spiders, and worms) and the soil microbial community, almond nutrient composition, crop yield, and profitability.

Many almond growers are hesitant about trying some of these regenerative practices for food safety reasons, Fenster told me. They worry that having any sort of vegetation between trees, much less letting animals wander through the orchard, might increase their exposure to *Salmonella* and *E. coli*. The regenerative perspective is just about the opposite. 'We think a healthy orchard floor with a diversity of plants, microbes, and insects will decompose the manure, and a diverse microbial community will keep the pathogens in check while also cycling nutrients more effectively,' says Fenster. He points to a study in Germany showing that children who grew up on farms with animals had more robust immune systems compared to those who grew up in urban areas without pets. In regenerative thinking, biodiversity gives the land a similarly robust immune system.

Because almonds are such a notoriously thirsty crop, many orchardists also worry — like conventional farmers everywhere

— that cover crops will siphon away the water needed by their cash crop. Again, regenerative agrarians hold the opposite view. The management of cover crops in arid California will definitely be different, Fenster says, from that in the rainier Midwest: the orchardist needs to either plant cover crops that naturally die down during the hottest months, mow them, or move animals through to chew and stomp them down. But cover crops, they argue, will actually make more water available to farmers. Cover crops not only reduce runoff during the rainy winter months, but they also build up the organic matter in the soil, which, in turn, boosts its water-holding capacity. They also diminish the amount of dust kicked up by harvesting, a boon to the lungs of everyone living nearby — already taxed by heavy pollution in California's Central Valley.

In the first year of the study, three of the regenerative orchardists were bringing different animals onto their property: broiler chickens, egg-laying chickens, and sheep. The orchardist welcoming the latter animal, Brian Paddock of Capay Hills Orchard in Esparto, California, is definitely having a good time with it. The sheep belong to his neighbour. Paddock is always eager to avoid the drudgery and fuel expense of mowing the cover crops in his orchard. He doesn't want to let them go to seed because that's a nitrogen-heavy transition, and he wants that nitrogen to stay in the ground for the almonds. Now, when the cover crops flower, he just announces to his wife that he's off to mow the orchard. Then he gets a beer and watches the sheep fertilise and mow at the same time.

Paddock tries to bring nature into his orchard in other ways, too. He stakes out owl boxes so that a larger owl population will snatch up the moles and voles. He's expanding his hedgerows — which provide habitat for native bees and other beneficial insects — and hoping to attract a fox for other animal pests like squirrels. He always knew he wanted to be organic because he never wanted to employ poisons or other dubious chemicals to control pests.

'You can go on YouTube and watch videos of people watching the nuclear blasts out in the desert,' Paddock told me. 'They didn't know at the time how dangerous that was. Why take the risk on these chemicals? My family and I live where we farm, so I knew I wanted to be organic.'

The results from Fenster's study — published in *Frontiers in Sustainable Food Systems* — echo those of Lundgren and LaCanne. Fenster found six times as much invertebrate biomass (insects, spiders, and worms!) at the regenerative orchards, compared to the conventional ones, as well as a significantly greater amount of diversity. The amount of crop damage by pests in conventional and regenerative orchards was the same, although for different reasons: the former minimised damage with chemicals and the latter's success was tied to their rising levels of invertebrate biomass and diversity. Fenster also recorded an average of 3.88 per cent organic matter in the regenerative orchards versus 2.39 per cent in the conventional ones — a significant difference, as studies show that for every 1 per cent increase in organic matter per acre, the land will hold an extra 19,000 to 25,000 gallons of water, depending on soil type. Sampling to a depth of sixty centimetres, Fenster's study found that the regenerative orchards hold 30 per cent more soil carbon, with his models indicating that regenerative orchards are building soil carbon while conventional orchards are losing soil carbon. Finally, Fenster found twice the microbial biomass in the soil of the regenerative orchards. While there was no difference in yield between the regenerative and conventional orchards, the regenerative orchards were twice as profitable due to the premium paid for their product.

The regenerative approach is spreading around the country, even among people who haven't come from and then broken with the conventional, commodity-production background — a background that has produced some of the world's leading spokespeople for regenerative agriculture. Through social media, Buz Kloot met an organic farmer in South Carolina who incorporates many regenerative practices on his land, many of which he just figured out on his own.

Even though he comes from a family with deep farming roots, Nat Bradford only began farming in 2012, when he was thirty-four — although not for lack of early interest. He was eager to go into farming when he was a student in his rural high school, but his agriculture teacher scoffed at the idea. 'He told me that unless I already had 2,000 acres of land and $2 million in equipment — already paid off — I'd always be in debt,' Bradford tells me. 'That's the dilemma of the modern farmer. I have friends in their twenties and thirties who don't want to do the same thing their fathers are doing, but they're stuck in the system. I'm trying to come up with a new model: making a living, making a positive impact on the land, and trying to figure out a way to help other farmers get into this new approach.'

Instead of going into farming when he got out of college, Bradford went into landscaping. He had a lot of ideas about improving on the conventional model of urban landscaping, in which plants are widely spaced on heavily mulched plots — what he calls mulch deserts — that require people to use Roundup to enforce their pristine weedlessness. 'Over 25 per cent of chemical pesticides are in backyards, schools, and municipal sites,' he says. 'I wanted to design landscapes that didn't necessitate Roundup in great expanses.' So his designs all incorporated various groundcovers that kept bare soils to an absolute minimum.

But Bradford still pined for farming. His grandfather had been a part-time farmer and made enough money doing it to send a son through law school and another son through medical school, and Bradford had 'fallen in love with produce early on'. He finally bought ten and a half acres that had been conventionally farmed in cotton, corn, and soybean monocultures, and began raising watermelons, collard greens, and okra. He committed early on to not using any chemicals and to bringing back the soil's health and fertility using only cover crops and manure. It took several years, but his results are impressive.

Key to Bradford's success is a carefully timed mix of cover crops that flower, go to seed, and die back just before it's time for him to start his market crop, which he plants directly into their matted, decomposing biomass. By the time his market crop is a foot high, those cover-crop seeds have already begun to grow again, enriching the soil, and attracting beneficial insects. Bradford is mimicking nature in two ways: he keeps a live root in the ground year-round, and he relies on the fact that soils are full of seeds that will sprout when conditions are right. Some seeds last in the soil for hundreds of years or more; in the soil of one dry lake bed, people have even found viable lotus seeds carbon-dated back 1,200 years.

'I'm trying to build a self-regenerating seed bank of cover crops in the soil,' Bradford says. 'I hate buying seeds and taking the time to replant them every year when I know they'll go to seed. That's what nature does! You don't have to replant a forest — you just walk away for fifteen years and the forest has regenerated, because the seed bank is there.'

Collards are Bradford's winter crop, grown on a single acre and swaddled in clovers and vetch. When the rains fall, the drops never strike bare ground, but splash onto the plants and then trickle down into the soil. (I'm an irreligious person, but I nonetheless think of this as God's drip-irrigation system.) The cover crops protect the soil from both rain and wind erosion, and the collards never get splashed with mud — a minor point, except to the chefs who buy Bradford's produce and say they hardly have to wash it. After the collards are finished, he puts in a crop of okra.

I spoke to Bradford in March 2019, when it was still the winter growing season, and he had been harvesting $22,000 worth of collards per acre since the previous December. When I spoke to him again in 2021, his crop yield had more than doubled to 13,500 pounds and $50,000 per acre, which he attributes to the continuing improvement of soil ecology. His okra production from the previous season had been likewise dazzling. Until production stopped because of a hurricane, he and his family were harvesting 22,000 pounds of okra per acre on each of their seven acres in production.

'That's close to four times the commercial yield for okra,' he says. 'I can't even imagine how many chemicals they'd need to use to match that. Organic farmers have to be honest in that conventional agriculture is our competitor. And with this kind of production, I can compete with them on yield, flavour, quality, soil health, and price.'

And watermelons! Bradford uses seeds that have been carefully passed down through his family for 170 years. Every year, the family went through the fields and identified the healthiest and most resilient plants with the most flavourful fruit and saved those seeds, which became the seed stock for the following year. Planted into his healthy soil and surrounded by cover crops, his watermelons are also going gangbusters. Bradford pointed to the cold, wet summer of 2013, when then governor Nikki Haley requested federal disaster aid in South Carolina after heavy rains devastated crops. The cucurbit industry — squash, cucumbers, and melons — suffered an outbreak of downy mildew. Many farmers lost their entire crop, but Bradford had over 100 per cent yield that year, as some of his vines surprised him with two huge melons, not just one.

Bradford knows that he is extraordinarily fortunate to inherit these seeds that have been grown for generations in the same area, interacting with the same soils, ecological community, and microclimate for so many years. His family saved seeds in hot, dry years as well as cold, wet ones. They likewise saved seeds from plants that thrived in years when there were insect onslaughts and disease. They grew the melons without the help of fertiliser and pesticides, so these plants know how to forage for nutrients and fight pests and disease without chemicals. Their genetics are a vast bag of tricks the plant uses to thrive, no matter how many challenges they face.

Most farmers in the United States — and other countries in the thrall of industrial agriculture — no longer have seeds that are adapted to their local conditions, much less seeds with the buff genetics to thrive in an organic system. They had them back in the days when all farmers used to save seeds from their best crop plants for the next season. They still had them in the last century, although to a lesser degree, when there were hundreds of regional seed companies that sold varieties that grew well for local farmers and pleased the palates of their families and customers. But the craftsmanship of seed saving was gradually lost among the vast majority of the nation's farmers, as industry took control of this most basic and precious agricultural resource.

Before that, almost all seeds came from plants that were open-pollinated. In a field of open-pollinated corn, for instance, pollen from one tasselling plant blows around to other tasselling plants and they fertilise each other in a marvellous swirl of cross-pollination and genetic possibility. Even if the field is planted with a variety for which farmers have carefully selected seed, resulting in big sweet ears and some other shared characteristics — height, maybe, or time of maturation or colour variation — there is still a massive amount of genetic differentiation from one plant to another.

In an open-pollinated field, those genetic differences offer farmers thousands and thousands of options for the future. Over the millennia, farmers exercised those options, creating distinct varieties in subregions around the world, with plants continually adapting to the desires of the various humans who grew them and to the places where they were planted. Each microregion, even each farmer, could boast a unique variety. And every year, the genetics of the variety would shift a bit as the environment and humans made their selections.

'When was America great?' Winona LaDuke, Native American activist and farmer, tossed the audience this question during her keynote address at a Portland conference called Organicology, where plant breeding and seed saving were on the agenda. 'America was great when we had 8,000 varieties of corn.'

In the early twentieth century, professional plant breeders began to shift this reliance on open-pollinated crops in favour of plants with much more limited genetic possibility. They developed hybrids, which were crop varieties bred from two inbred parents, themselves created by generations of self-pollination. Seed companies were able to distinguish themselves and build their businesses with their own proprietary hybrids, guaranteeing farmers that they would get exactly the same result with every planting. Those uniform, widget-like fruits and vegetables fit well with the needs of the nascent food industry, ensuring farmers a market. But farmers could not save seeds from the hybrids, because the next generation would be genetically unstable and revert back to the qualities of the inbred parents — not, usually, what the farmer wanted. They had to buy new seeds every year.

'With open-pollinated crops, there is enough genetic diversity from one season to the next that the crop continues to evolve and adapt to environmental conditions,' says Micaela Colley, program director of the Organic Seed Alliance. 'But the hybrid has a different goal: it's the fixed endpoint of a variety with certain qualities and traits. It's not meant to evolve.'

After the advent of biotechnology in the late twentieth century and the patenting of living things including seeds, giant chemical companies began gobbling up those regional seed companies. Now, four massive monopolies — Corteva (a merger of Dow and DuPont), ChemChina (which swallowed Syngenta), Bayer (which swallowed Monsanto), and BASF

— are estimated to control over 60 per cent of the world's seeds. Because these chemical conglomerates are vacuuming up smaller seed companies to make a healthy profit, not to promote healthy agriculture or landscapes or people, they have 'retired' thousands of the smaller seed companies' proprietary hybrids — meaning, they no longer breed the parent lines that create the hybrids and grow them out for distribution. In the year 2000 alone, more than 2,000 hybrids disappeared from the marketplace when Seminis — at the time, the world's largest vegetable seed company — bought several smaller companies. Instead of continuing to offer seeds for thousands of crop varieties, it is more profitable for the conglomerates to prune their portfolios down to a smaller offering of seeds with big markets: either ones that grow reasonably well everywhere or ones that are tailored for the major commercial growing areas — think California for tomatoes and the Midwest for corn. The conglomerates then target some of the most profitable hybrids for genetic tinkering, in almost all cases so that they can increase farmers' reliance on their chemicals. According to Bill Freese, science policy analyst for the Center for Food Safety, GMO varieties comprised 93 per cent of the US acreage planted to five major crops — corn, soybeans, cotton, canola, and sugar beets — in 2020. Ninety-eight per cent of these GMOs are engineered for resistance to glyphosate and other herbicides. Most of the GMO cotton and corn is also engineered for insect resistance.

In the United States and other countries where seed-saving

is almost a lost art, farmers now have a hard time finding the older varieties they used to rely on. Farmers growing commodities like corn, soy, and canola often can't buy anything but GMO seeds precoated with the conglomerates' pesticides. If they're trying to farm ecologically, they find these seeds problematic. If they're planting cover crops to attract beneficial insects, they often find their hopes thwarted, as the insecticides that coat these crop seeds can taint the nectar, pollen, and tissues of the cover crops and leach into the water table.

'Are we poisoning the well for these beneficial insects?' asks Mike Bredeson, Jonathan Lundgren's former grad student, who's studying this. 'Yes, it appears we are.'

Lundgren's research casts doubt on whether these costly pesticide-coated seeds are actually helping farmers with their pest problems, and research from University of Idaho biologist Mary Ridout suggests the same. Ridout compared corn seeds treated with fungicides to untreated corn seed, putting one cup of each into jars partially filled with water. Four weeks later, there was far more fungal growth — especially among three or four species of fungus — on the treated seeds than the untreated ones, where there was greater fungal diversity. The fungicide hadn't eliminated fungal growth, she concluded; it had only altered the diversity of the fungal community and eliminated the competition that might have held those three or four more aggressive fungi in check.

Even the commercially produced seeds that aren't coated with pesticides are a feeble simulacrum of the kind of robust

A white-winged dove (left), a Gila woodpecker (right), and bees share fruit, pollen, and nectar from a saguaro cactus in the Sonoran Desert in North America. The woodpeckers peck holes in the cactus for their nests, which can later be reused by owls, wrens, and other species. BARBARA CARROLL/GETTY IMAGES

More than 600 years old, this red cedar lives in Kokanee Glacier Provincial Park, Canada. Forest ecologist Suzanne Simard calls these older, larger trees 'Mother Trees', and studies their oversized role in forest health. BRENDAN GEORGE KO

Suzanne Simard in Nelson, British Columbia, digging up the surface soil to reveal fungal strands growing around Douglas-fir roots. BRENDAN GEORGE KO

Suzanne Simard examines mycorrhizae — a composite structure formed by fungi and roots. BRENDAN GEORGE KO

A cheating bumblebee, caught in the act! Here, the *Bombus bifarius* acquires nectar from a Sierra fumewort in California by entering a hole in the bottom of the blossom — probably created by an even larger *Bombus occidentalis* — instead of entering through the open blossom. Biologists call this behaviour 'cheating' because the plants are no longer being pollinated in return for the nectar they provide. CAITLIN WINTERBOTTOM

In some locations, this bee orchid lures male bees with blooms that both look and smell like female bees; as they attempt to copulate, the male bees pollinate the flowers. In the plant's northerly range, such as here in Dorset, England, the orchids self-pollinate. BOB GIBBONS/SCIENCE PHOTO LIBRARY

**Above:** Hundreds of miles off the coast of western Canada, a lion's mane jellyfish provides refuge for a juvenile prowfish as blue sharks circle nearby.
IAN MCALLISTER

**Right:** This bobtail squid in the Philippines uses its partnership with a bacterium to create luminescence.
TODD BRETL

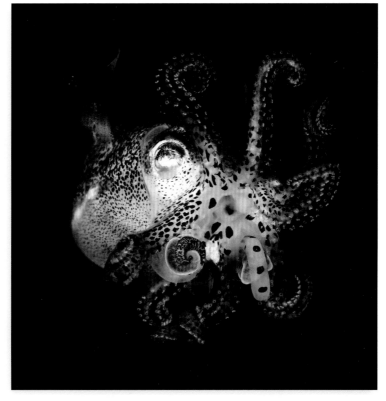

Born to Russian aristocrats in 1842, Peter Kropotkin became one of the world's leading anarchists and one of its most famous scientists, arguing that cooperation within species helps them survive. In his bestselling book *Mutual Aid: A Factor of Evolution*, he wrote that if we ask nature, "'Who are the fittest: those who are continually at war with each other, or those who support one another?' we at once see that those animals which acquire habits of mutual aid are undoubtedly the fittest.' PUBLIC DOMAIN

Near the Great Bear Rainforest, Canada, humpback whales engage in a cooperative hunting strategy for herring called bubble-net feeding. The whales, in coordination with each other, blow bubbles around a school of fish from below, driving them to the surface in a tight group, and then feed by coming up from below. IAN MCALLISTER

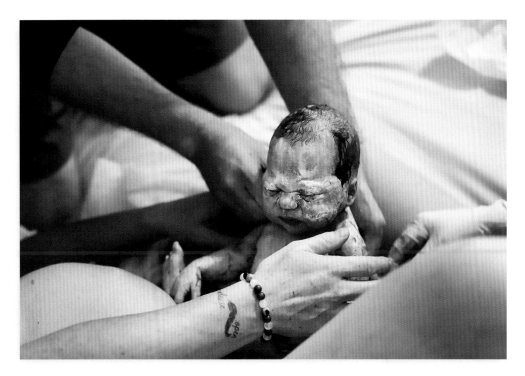

A newborn baby covered with a nurturing slick of microbes from his mother's birth canal.
LILIANA LEAHY

The remains of this pink salmon enrich the Alaskan soil with nitrogen carried from deep in the ocean. CHRISTOPHER MILLER

Brain corals and other reef-building corals have a mutualistic relationship with the photosynthetic algae that live in the tiny coral animal's soft tissues. The coral provides protection to the algae and feeds it with its own nitrogen waste, while the algae in turn feeds the coral with the carbon sugars it gleans from photosynthesis. NORBERT WU/MINDEN PICTURES

*Geotrichum candidum* is a member of the human microbiome, notably associated with skin, spit, and faeces. It is also common in soil and is used to make cheeses such as Camembert, Saint-Nectaire, and Reblochon. M OEGGERLI (MICRONAUT 2019) WITH ANDREA AND ADRIAN EGLI, CLINICAL MICROBIOLOGY AND PATHOLOGY, UNIVERSITY HOSPITAL OF BASEL, AND BIOEM LAB, BIOZENTRUM, UNIVERSITY OF BASEL

**Above:** A common bonnet mushroom and a holly seedling growing in a symbiotic relationship in the New Forest National Park, England. GUY EDWARDES/MINDEN PICTURES

**Right:** At Capay Hills Orchard in Esparto, California, sheep are a highly valued partner — they remove annual cover crops and weeds, reduce tractor time, and provide fresh fertiliser. BRIAN PADDOCK/CAPAY HILLS ORCHARD

**Above:** This 1980 image shows the denuded conditions along Maggie Creek (on land owned by Nevada Gold Mines) near Elko, Nevada, before ranchers changed their grazing practices and the beaver returned. BLM, ELKO DISTRICT

**Below:** The same spot as above, rejuvenated by changed ranching practices. CAROL EVANS

**Right:** At the Finca Irlanda organic coffee plantation near Tapachula, Mexico, coffee plants grow among shade trees and are surrounded by a wealth of biodiversity. BROWN W. CANNON III/ALAMY

**Below:** A female phorid fly attacks fire ants in a captive study conducted by Dr John Abbott at the Brackenridge Field Laboratory, Travis County, Texas. JOHN ABBOTT/MINDEN PICTURES

**Above:** Eric Horvath and Claire Smith on their property near the north fork of Oregon's Beaver Creek, where logs were dropped into the creek to create more complexity and better habitat for salmon. PAUL ENGELMEYER/MIDCOAST WATERSHEDS COUNCIL

**Left:** A difference of many degrees on a South Carolina plot: on the right, residue from cover crops continues to keep the soil fifteen degrees cooler than the bare soil immediately to the left. Cooler soil not only holds moisture, but also provides a welcoming habitat for soil microbes and beneficial insects. BUZ KLOOT

**Above:** The wild ancestors of modern tomatoes, like these from Peru, may have traits that plant breeders can tap to improve new varieties. TONI ANZENBERGER/ REDUX

**Right:** University of Wisconsin–Madison agronomy chair and plant breeder Bill Tracy inspects a cob of Who Gets Kissed? organic sweet corn, dried for seed. JOHN HART/ *WISCONSIN STATE JOURNAL*

A stunning array of plate corals crowd Opal Reef in the Great Barrier Reef, Australia.
DAVID DOUBILET

Singapore's Gardens by the Bay is a nature park spanning 250 acres of reclaimed land in central Singapore, adjacent to the Marina Reservoir. SHAN SHIHAN/GETTY IMAGES

**Above:** Sponge Park in New York City uses specially chosen plants to extract heavy metals and biological toxins from contaminated stormwater, protecting the nearby Gowanus Canal in Brooklyn. DLANDSTUDIO PIIC

**Below:** Beavers shape environments more than any animal other than humans. As they dam streams and rivers, they create vital habitat for birds, fish, and other wildlife. Beaver dams also slow the passage of water and can even raise the water table of the surrounding landscape. CAROL 'CAZ' ZYVATKAUSKAS

A pair of red-billed oxpeckers search for ticks, fleas, and lice on an impala in Mpumalanga, South Africa. The relationship is mutually beneficial for both species: the impala gets rid of harmful parasites, and the oxpecker gets a tasty meal. HEINI WEHRLE/MINDEN PICTURES

seeds Nat Bradford inherited. The adage among plant breeders is that plants should be bred in the environment of intended use, but the environment where most commercial crop varieties are bred is high on chemicals and low on diversity. They're bred with industrial expectations: that they will be goosed with chemical fertilisers, protected from insects with applied pesticides, saved from disease with applied fungicides, and relieved of competition with weeds by herbicides. Because of the environment in which they're bred and raised, these varieties may lack the genetic wherewithal to thrive in an organic system — or even one in which farmers are either trying to save money or save the rest of nature by applying fewer chemicals.

Like humans and every other complex organism, plants have a dynamic microbiota. When we look at a plant, we see only the visible part of an entire ecosystem that includes bacterial, fungal, and other partners in the soil as well as microscopic organisms tucked into the plant's tissues. Each plant's relationship with these partners begins in the seed, which hosts a small population of microorganisms. According to research looking at the domestication gradient from corn's most ancient ancestor, teosinte — a grass from southern Mexico — to modern corn, each iteration seems to host the same core microbiota in their seeds as the wild varieties.

Scientists are still working on what those core symbionts do for the seed as it germinates, which Ridout says is the most vulnerable time in the life of the plant — it's fatally susceptible to lack of water, nutrients, hard soil, harsh environments, and pathogens. So far, research suggests that the symbionts confer better germination, faster growth, and increased biomass both above ground and below. In addition, they seem to act as the plant's gatekeeper, protecting it against pathogens and helping it form connections with microbial partners in the outside world.

As the plant continues to grow, it needs a genetic toolkit to fight pests and disease on its own; it also has to know how to muster a handshake with these new microscopic partners to feed and defend the ecosystem. Microbes boost the growing plant's defence in at least three ways: they can produce chemicals that protect plants, they can outcompete hostile organisms around the plant, and they can trigger what's called an induced systemic resistance, where even the very presence of a beneficial microbe can put the plant on alert and ready to mount a defence against pathogens if needed.

Scientists are just starting to probe whether and how much plants have lost their ability to work with their microbial partners as a consequence of centuries of domestication and plant breeding. In 2019, Lori Hoagland — a Purdue University plant breeder and soil microbiologist who is trying to breed tomatoes with these ecosystem considerations in mind — travelled to Colombia, where tomatoes originated and still grow wild. She hiked into the mountains to collect root tissue from the wild tomatoes as well as the soil surrounding them. Back in her greenhouse, she's studying how the microbes in this soil help these ancestor plants survive stress from pathogens and drought, compared to how much they help modern plants. 'If these microbiomes turn out to be important, we could try to breed these relationships into our modern varieties using traditional breeding techniques [i.e. not GMO],' Hoagland tells me. 'We might also be able to manage our soils in agricultural systems to be sure that these types of microbes are present to help our modern tomatoes.'

It's possible that modern plant breeding has basically turned most crop plants into idiots — my hyperbole, not Hoagland's — that no longer have the wherewithal to seek out their friends and thrive. Instead, the farmers who plant these seeds are forced to buy a multitude of costly products to stand in for the services these ancient mutualisms would otherwise provide for free. No wonder the USDA forecasted that debt among the nation's two million farmers would soar to $441.7 billion in 2021. No wonder farmers are among the populations most likely to commit suicide.

'It's huge money for a few,' says Idaho organic potato breeder Clyde Bragg. He was once a vice president with Green Giant in charge of developing new varieties and found that the best potatoes came from organic seed, but he couldn't convince the Giant to convert to organic. So he quit and began breeding and raising his own potatoes, saving the $800–$1,000 in chemicals per acre that conventional potato farmers pay. 'We're losing farmers, but the chemical companies keep getting bigger. I don't know one person in the chemical industry who's lost their livelihood, but I know lots of people who have lost their farm.'

But just as there is a grassroots movement to restore ecological wisdom to farming, there is a complementary movement underway to restore the lost art of plant breeding and seed saving as it's been practised for thousands of years.

Organic farmers realised the need for such a movement back in the 1980s. They were making do with a handful of commercial hybrids that worked for them, as well as with the so-called heirloom seeds — varieties that had been around for at least fifty years and are either open-pollinated or self-pollinated — that began to appear in seed catalogues like Johnny's Selected Seeds in the 1970s. They preferred the heirlooms for a number of reasons. Many found hybrids philosophically objectionable and wanted independence from the big seed companies, which increasingly focused on creating varieties that suited the standards of agribusiness — crops that ripened at the same time, transported better, and lasted longer, but often lost the flavour, texture, and beautiful colours of the older varieties.

But the heirlooms also weren't a perfect solution for organic farmers who actually wanted to make a living. 'The heirlooms tasted great, but they often cracked and didn't ship well,' recalls John Navazio, who became an organic farmer in the 1970s and then went back to school in the 1980s to learn traditional plant breeding. For a while, he was Organic Seed Alliance's senior scientist and the organic seed extension specialist for

Washington State University, but he is now working at Johnny's Selected Seeds in Maine. 'You could hardly even pack those heirlooms on a truck and take them to town to sell them. Conventional tomatoes didn't crack, but they didn't taste good either. I started to wonder if we could put those characteristics together and soon realised there was no one breeding varieties for the farmer marketing high-quality organic produce on a local scale.'

By the early 1980s, organic farmers had begun to conference about their need for improved varieties and well-produced seed. Frank Morton — now the legendary plant breeder and seed seller behind Wild Garden Seed in Philomath, Oregon — recalls going to a meeting in 1984 where a molecular biologist known as Mushroom (named, Morton says, because he claimed 'my friends keep me in the dark and feed me shit') stood up and announced, 'If you grow organic crops, you need organic seeds. Those seeds don't exist, and we have to create them.'

'It blew my mind,' says Morton, who was growing organic lettuce for upscale restaurants back then and now sells seeds for the more than eighty new organic varieties he's created. 'But my real epiphany came when I realised that if I kept saving my seeds, I'd find accidental crosses and could grow them out and get a genetic rainbow.'

The Organic Seed Alliance (OSA) was organised in 2003 and now includes both farmers interested in developing their own varieties — pushing the evolution of existing varieties so that they fare better against pests and climate challenges — as well as several professional plant breeders from public universities. Because of cuts in government funding to these universities, there aren't many of these breeders left, and not all are dedicated to organic. But the ones working with the OSA believe that neither agricultural abundance nor a healthy planet are going to come from the genetic-manipulation tools of biotechnology, which change varieties in the narrowest of ways, one or two genes at a time. Instead, they favour making selections from open-pollinated plants grown organically, a practice that engages the entire genomes of the plants.

'If we want to have a truly sustainable future,' says Michael Mazourek, a plant breeder from Cornell University who's an OSA stalwart, 'we need plants that are robust and resilient intrinsically, from their genetics. There are tens of thousands of genes in two plants, and most plants also have different forms of the genes, called alleles. When we do a cross-pollination, we're working with all that diversity, well beyond what we will be able to define and characterise for a long time. It's underappreciated how powerful that is.'

The OSA has developed a model for breeding, similar to Lundgren's model for agricultural research. Called participatory plant breeding, it takes place on farms with the active engagement of farmers, often to address a need voiced by growers in that region. 'It's the way plant breeding should always work,' Mazourek says. 'When most breeders develop a variety, the next step is to get farmers to try it. But with participatory plant breeding, the beneficiaries are the co-developers of that variety and as soon as it's ready to be used, it's already been adopted by them.'

By participating in the breeding process — or by taking one of the many classes the OSA and some of the public plant breeders offer — farmers also learn how to make selections and save seeds themselves. This wonderfully democratising development allows farmers greater freedom from the chemical conglomerates and also helps them craft — along with eaters, chefs, and others — stronger local food systems.

An example: the OSA is a member of a partnership called the Northern Organic Vegetable Improvement Collaborative (NOVIC), which works with farmers across the northern tier of the United States to breed crop varieties that meet their particular needs. An organic farmer near the Twin Cities named Martin Diffley mentioned a concern to them. He had been raising an older commercial sweet-corn hybrid called Temptation for many years, and a big chunk of his income came from selling it to a public hungry for local corn every summer. No other variety grew well in this part of Minnesota, where spring arrives late and soils remain cool. Most sweet-corn seeds don't germinate quickly enough there to shoot out of the soil and start photosynthesising; instead, they break down and microbes start to chew on them. But Temptation performed like a champion and was a foundation of his business.

However, Monsanto now owned Temptation and announced that it was coming out with a GMO version. Past experience suggested that Monsanto would discontinue the original, and Diffley worried that he wouldn't have a usable variety anymore, as certified organic farmers can't use GMOs. So NOVIC turned to public plant breeder and OSA ally Bill Tracy, at the University of Wisconsin, who had been breeding sweet corn for years, and asked him to look through his vast collection of varieties for traits that might work. Tracy's grad student Jared Zystro — now the OSA's assistant director of research and education — planted Tracy's offerings on over a hundred breeding plots on Diffley's land, and the farmer himself selected the traits he wanted: vigorous plants that tolerated cool soil and resisted disease, tasted great, and had large ears with husks tightly wrapped at the end (which protects the ear from insects). Tracy then grew out Diffley's picks at his winter nursery in Chile, and the best of these were planted the next year back on Diffley's land. This process went on for four more years, and in 2014 NOVIC released Who Gets Kissed? — an open-pollinated, organic sweet corn for farmers with cold, late springs.

The name harkens back to the days when 'regional farming' was an oxymoron, because all farmers were primarily growing for their families and local communities. 'Back in the old days when the corn would ripen, communities would come together and have a husking party,' explains Micaela Colley. 'If you found an ear with a red kernel, you got to pick who you wanted to kiss in the circle. Naming this new variety Who Gets Kissed? was our way of saying that it's not the end of the world if you find a red kernel! We're trained to want uniformity, to want our crops to be identical little tin soldiers. But that's part of the reason we don't have the resilience of biological diversity in our food system.'

Then the OSA continued the evolution of sweet corn for a different group of northern farmers. Colley and others from the OSA were meeting with a food co-op in Washington State about how they might help local farmers provide fresh produce year-round. It turned out that farmers in the northwest had also been using Temptation successfully and were worried that it would disappear, but Who Gets Kissed? wouldn't cut it for them — Minnesota and the Pacific Northwest share cool, late springs, but the summers in Minnesota are much hotter and the farmers can bring Who Gets Kissed? to market in a few months. In the Pacific Northwest, they might not be able to do so until Thanksgiving. So with a grant from the food co-op, the OSA and Bill Tracy began trialling a new assortment of sweet-corn varieties with Nash Huber, a local organic farmer who markets directly to Seattle and sells to wholesalers from Portland to British Columbia, and other organic farmers. The selection process went through the same stages as Who Gets Kissed?, although this time the OSA and the farmers were looking for the same big-eared, flavourful, open-pollinated, organic sweet corn — tolerant of cool soils and late springs and resistant to pests and disease — as well as one that matured early enough, even in a cool climate, to harvest in late summer.

I've attended two OSA conferences — they take place

every other year — and I'm always dazzled by the range of people I see there. There are, of course, the senior statesmen of farming and plant breeding, and theirs are the faces called up by the alarming news bulletins about the ageing of our agrarian population and agrarian scientists — white, male, and over sixty. But at the OSA and other innovative ag-related conferences, young people, women, and people of colour swamp their demographic. I keep waiting for the news bulletins to catch up with what I'm seeing.

At my first morning at an OSA conference, I plopped my-self at a table with two women and asked about their con-nection to farming and seeds. One was in her early thirties, a graduate from a prominent liberal-arts college, and had been farming — and was now starting to produce seeds as well as crops — since she graduated. She works hard, she told me, but it's her own business and she loves it. Her friends who are employed on Wall Street and in law firms envy her. The other woman was older, maybe in her forties. She was the third gen-eration in her family to farm; she also confessed to working hard but loving it. They and many of the people around me had the eagerness and zeal of converts to a new religion.

There's not as much to love about conventional farming, with its rigid schedule of chemicals and tillage while — over your shoulder — you see the greater landscape's biodiversity dwindle to dust. That's what Mike Bredeson told me, back in Roger's cornfield, back when he was still a PhD candidate. We started off talking about the research he was doing there, how he hoped it might someday lead to greater plant and insect diversity in America's cornfields. He praised the special role of sunflowers as part of a cover-crop mix: they have extrafloral nectaries (meaning, sources of nectar that are outside the flower, along the stems and leaves) and feed beneficial insects whether they're blooming or not. Then he bent down and pointed to a white dot bobbling alongside a corn leaf, like a balloon tugging on a tiny string. A green lacewing's egg, he told me, which the mother tethers to the plant with silk to protect it from ants, which are trying to protect their aphid livestock from predation. Once the egg hatches, the lacewing larvae — called aphid lions or aphid wolves and resembling tiny brown alligators — will gobble thousands of aphids and other crop pests.

And then he said something really surprising. After he finishes his PhD and he and his wife settle down somewhere, he wants to be a farmer. 'The farmers I'm working with give me hope,' he says. 'They have the most interesting and important job on the face of the Earth, and that's to be [a] steward of the land.'

Before meeting Lundgren, Bredeson didn't want anything to do with farming. He had grown up on a conventional farm himself and wanted to get as far away from that life as possible. He had been pre-med and, after his sophomore year, decided a summer job doing some kind of biological research would make him a more appealing applicant to medical school. He heard that a local entomologist needed summer help, and he wound up getting a job with Lundgren.

His first day changed the course of his life. Lundgren started him off doing the kind of work I had observed, sucking up insects from the soil. 'I had lived on a farm for nineteen years and never once got down on my knees and pushed back the soil surface to see what was happening there,' Bredeson says. 'Jeez, I never saw anything so interesting in my life! There are alliances and wars being waged beneath our feet on a daily basis. At that moment, I thought, humans are way less interesting than this.'

## Chapter 6

# I'll Take My Coffee with Birds

Ivette Perfecto raps her knuckles on the enormous sope tree shading the nearby coffee plants, as vigorously as if she is knocking on someone's front door and expecting them on the doorstep pronto. The tree's inhabitants quickly muster the predicted response: dozens of small brown *Azteca* ants rage out, ready to inflict damage on the intruder. Perfecto steps back and peers closely to see if the next act in this ongoing ecological drama will unfold. Sure enough, a tiny fly begins to buzz the ants, drawn by the alarm pheromones they emit when disturbed.

'There is the phorid fly!' she cries out gleefully, gesturing me closer. 'And here come more of them.'

We lean in to see if one of the flies will settle on an ant head and begin to lay eggs. If it does, the larva will develop inside the head and nourish itself on the brain, eventually causing the head to detach and tumble to the ground, releasing a new fly to torment still more ants. But the pattern of frenetic activity among these particular flies never quite matches that of the ants; they just fly in erratic circles. 'They're confused,' Perfecto says finally. 'Different phorids attack different types of ants to lay their eggs, and this isn't the right type of ant for these flies. They're very specific.'

I suddenly feel a sharp stab at the back of my neck. I had been cautioned by my hosts not to let branches and leaves brush up against me if I want to avoid intimate contact with the *Azteca*. Perfecto and John Vandermeer are ecologists at the University of Michigan who've long been probing the interactions among living things on this organic coffee plantation near Mexico's southern border with Guatemala — they are marital partners and my guides on this outing. The *Azteca* avoid soil — 'they don't like to get their tootsies dirty,' Vandermeer quipped — and navigate the plantation via trunks, branches, leaves, twigs on the ground, and the huge roots of the sope tree, which rise out of the soil like knee-high, serpentine walls. I realise that I must have let part of the tree touch me as we were talking. 'They don't sting,' Vandermeer tells me, looking slightly amused. 'They bite and then turn around and inject some poison into the wound.'

I retreat to watch Perfecto and Vandermeer do their work in the doughnut of greenery around the sope. I stand far enough from trees and coffee plants and fallen branches to be safe from the fierce *Azteca*, the scourge of the plantation's workers. I eye the ground nervously, as one of the first stories I heard when I arrived was of a former student returning from a day of field work, who dropped her knapsack on the field-station floor and recoiled, horrified, as two coral snakes oozed out. So I poke my walking stick in the piles of leaves nearby to make sure I don't step on a coral snake — we carry the sticks to ward off any of the farmworkers' dogs who might sneak up for a nip, as well as to stay upright on the paths snotted with orange mud after a recent downpour. I move every few minutes to avoid the chorus of mosquitoes that circle whenever one stands still for too long. And I watch as Perfecto and Vandermeer check the distribution of the many other players in the plantation's intricate ecological drama that they have been parsing out and documenting for the last twenty-five years. They begin speaking Spanish, the default language for them and all their students at the field station — and of course, the language of the farmworkers and the plantation owner, all of whom contribute bits of information that further their research. Then they remember that their monolingual guest is hunkered in the background, struggling to keep up with what's going on. They switch to English, Vandermeer groaning, 'This is going to be hard.'

A shocking comment several years ago by someone I regard as an environmental and scientific hero led me to the couple's work. On a reporting assignment from *Discover* magazine, I had attended a conference at the University of California, Berkeley, where scientists who conduct research in the US national parks discussed their work as well as the threats to those protected lands and the creatures that live there. The keynote speaker was the venerable biologist and Pulitzer

Prize–winning author Edward O. Wilson, who has famously declared that it is 'the little things that run the Earth' and has himself described 450 new ant species. Wilson discussed the devastating extinction rates among species worldwide and pitched his 'nature needs half' proposal to increase the world's protected areas from the current 15 per cent of the world's land mass to 50 per cent.

'The only way to save the rest of life, if we want to save the rest of life — and heavens to Betsy, we do! — is to increase the area of protected and inviolable habitat around the world to a safe level,' he implored the audience.

Wilson noted that the world was turning green, but only a 'pastel green'. That even among people who worry about the health of the planet, there is a misguided focus solely on protecting the physical environment — the climate, the air, the amount of fresh water and arable soil. The Earth's biodiversity and the living environments where that biodiversity is concentrated receive comparatively little attention, he said, even though the health of the physical environment and Earth's living things depend upon each other. If we focus on only one, we'll lose them both.

I sat in adoring attention, inwardly applauding every word.

But later in his speech, Wilson started talking about global trends that can help humans make room for the rest of nature. We already produce some 2,800 calories daily for every human, he said approvingly, but there are problems with transport and the poor quality of what he called artisanal production. 'We can fix that,' Wilson said. 'The present-day agriculture is still, basically, around the world, Neolithic. There are ways, particularly aided by the digital revolution, of greatly increasing, with dry-land production and so on, output.'

'Did he just endorse industrial agriculture?' a shocked young scientist sitting next to me whispered. I was also shocked. Who am I to question the wisdom of someone who's won more than 100 scientific prizes and written foundational books about the natural world, I thought. But then again, Wilson seemed to be saying something that was profoundly wrong.

I came to understand that there is a sharp disagreement among conservationists about how to conduct agriculture so that it provides for human needs without displacing or destroying the rest of nature in catastrophic ways. The argument boils down to 'sharing versus sparing'. Sharing proposes working lands imbued with nature. It comprises all the forms of agriculture that include nature-friendly practices — farming that eschews chemicals and monocultures and tilling, that incorporates woodlots and hedgerows and prairie strips and compost and such; farming that tries to support and include the rest of nature instead of eradicating it or disabling many of the relationships that make it function.

But some conservationists see any kind of farming as an assault on nature, in which even a benign disturbance is too much. They favour the sparing approach, in which agriculture operates on a small footprint with the full arsenal of modern chemicals and technology to maximise yields. That smaller footprint will allow, they believe, the creation of much larger wildlife preserves with little-to-no human impact. The sparers assume that nature-friendly farms can't produce enough food for the burgeoning population and will only lead to more and more wild landscapes being converted to farms.

While some nature-lovers have always looked upon agriculture as nature's antithesis, the sparing perspective may have been first articulated by Norman Borlaug, considered to be the father of the so-called Green Revolution. Trained as a plant pathologist, he began working in the 1940s for the Rockefeller Foundation in Mexico to create improved wheat varieties, in advance of what were projected to be mass famines around the world. He first turned his attention to breeding new varieties adapted for the tropics that were resistant to a fungal disease called rust. But his greatest impact on worldwide food production came when he began breeding varieties of wheat that could take advantage of a relatively new invention: the synthetic nitrogen invented in 1909 by German chemist Fritz Haber, which had applications in both munitions and agriculture.

By the time Borlaug was working on his wheat varieties after World War II, American chemical companies had an abundant supply of nitrogen that hadn't been turned into bombs, and they were eager to redirect the glut to agriculture. Like many other crops, wheat grew like gangbusters after an application of the nitrogen fertiliser, but the heads became so heavy that the plants flopped over. So Borlaug bred a shorter, sturdier variety of wheat that would allow the enormous yield gains from nitrogen fertiliser, but produce plants that could hold up their heads. Aside from believing that this work would stave off world hunger, Borlaug also said that 'by producing more food per unit of cultivated area, more land would be available for other uses, including recreation and wildlife'.

According to an article by author Fred Pearce in *Yale Environment 360* summarising the sparing versus sharing debate, some land has been saved by industrial agriculture, but far less than Borlaug anticipated. One of the scientists who promotes sparing admits that savings 'were far less than predicted by Borlaug, in the region of 20 million hectares rather than 560 million ... [and that] the higher yields were used primarily to produce more, cheaper food, not to spare the land for nature'. Conversion to industrial agriculture doesn't even necessarily mean more cheap food: in Brazil, for instance, Amazon rainforest and grasslands were converted to industrial agriculture, but the resulting produce did not feed Brazilians. Instead, it was exported to further enrich a small group of wealthy landowners.

The land-sharing conservationists on the other side of the debate aren't even that impressed with those twenty million hectares. These protected lands are mere fragments within a human-dominated landscape — usually agriculture, which covers 37 per cent of our land mass — and wildlife usually misses the memo that they're supposed to stay in the protected areas. The *Yale E360* article quoted biologist Stephen Kearney, who conducted a study of biodiversity in Australia, saying that 'it is not enough to just place land in a protected area and then walk away ... simply reserving land will remove all threats to very few species — only 3 per cent in fact.' As Pearce points out, even as the area dedicated to protection has grown significantly in the last few years, wild populations of vertebrates have nonetheless decreased 60 per cent compared to a half-century ago. Germany has seen a 78 per cent decline in insect biomass, despite an overall increase in that country's protected areas.

Many people also question some of the basic assumptions behind the sparing-sharing debate. One such assumption is that we must keep increasing agricultural productivity to feed a growing world. But according to the Food and Agriculture Organization of the United Nations, we already grow enough to feed ten billion people, which is one estimate of the world's peak population. A third of that production goes to waste, and another third feeds automobiles and CAFOs — Concentrated Animal Feeding Operations are where animals are divorced from their natural surroundings, crammed into very small spaces, and often fed things they never evolved to eat. The massive amount of food produced by industrial agriculture rarely reaches the billion people who are hungry, not because there isn't enough food, but because it's too expensive or just isn't locally available. And the farmers who are on the industrial-production treadmill suffer, too: the problem for them is often overproduction, which results in lower prices despite all their hard work.

Sparing also ignores the massive environmental consequences of industrial agriculture — the collateral damage to
the immediate landscape and beyond. As Vandermeer fulmi-
nated in a 2011 blog post about the sparing versus sharing
debate, 'Just how much more of the world's oceans are we
ready to sacrifice to dead zones? How many more aquifers will
we permit to be poisoned with pesticides? How much more
soil will we allow to be washed away? While agroecological
techniques are also implicated in some negative consequences,
these are trivial compared to the massive consequences
generated by the industrial system.'

I found Perfecto and Vandermeer not because of this blog
post, but because several scientists whom I contacted after I
heard Wilson's speech told me that the couple — along with
colleague Angus Wright — wrote the book on this subject
of land sharing. Their 2009 classic *Nature's Matrix: Linking
Agriculture, Conservation, and Food Sovereignty* argues that
preserves are definitely needed by some species that can't
tolerate any degree of human disturbance, but that the best
way to promote the survival of the vast number of species is
to improve the overall landscape — the matrix — around wild
fragments. Preserves are like islands. If they are surrounded by
chemical-drenched, heavily tilled monocultures — even the
tilled monocultures of the industrial organic enterprises —
they are like islands in a hostile sea. When I first called them,
Vandermeer said that most people tend to think of biodiversity
as things that look something like us — things with eyes that

we can look into, from birds to lions — but that microbes and insects comprise the vast majority of living things. They are the foundations of the matrix, and they are destroyed by intensification. Land sparing doesn't help them, he told me, but 'it is a winner for Monsanto, Syngenta, ADM, and the other giants.'

The land-sparing bias among many conservationists — perhaps stronger in the past than it is now — is one of the reasons Perfecto and Vandermeer wound up spending the bulk of their research career at Finca Irlanda in mountainous southern Mexico. Perfecto had never bought into the sparing philosophy to begin with. She had grown up in Puerto Rico, loving nature but also appreciating the small-scale agriculture of her neighbours.

'I was convinced agriculture didn't have to have a negative effect on biodiversity,' she told me as I interviewed the couple back in the Finca Irlanda field station, a humble string of rooms made beautiful by the pink bougainvillea draped over the entranceway and the dense greenery of the surrounding forest. 'I got that growing up in the tropics, where there wasn't a neat distinction between forest and farms.'

Both Perfecto and Vandermeer were interested in agriculture — she met him while working with his students on experiments with intercropping in a Michigan botanical garden — and they wanted to do research in the tropics, where the bulk of the world's biodiversity as well as the bulk of its poverty is concentrated. In the early 1980s, they spent three summers in Costa Rica teaching Latin American students about managed ecosystems. It was a time of great transition in Latin American agriculture, especially in coffee. Coffee is an understorey plant native to Africa — discovered in Ethiopia, although scholars think its beans were first roasted in fifteenth-century Yemen to make a beverage that allowed Sufi monks to stay up all night and pray — and it has been grown in Latin America since the late eighteenth century. The traditional approach was either to plant coffee into existing forest or, in unforested areas, to plant leguminous, nitrogen-fixing shade trees like the sope, as well as fruit trees, along with the coffee. Both kinds of farms harboured a considerable amount of biodiversity, as the larger trees provided food and habitat for a range of other living things.

But in the 1980s, international lending agencies had started to pressure countries in northern Latin America to modernise their coffee production to help service their debts. The United States, through its Agency for International Development (USAID), provided $81 million to finance this modernisation. The program offered farmers incentives to abandon traditional coffee varieties for new, higher-yielding ones that could grow without shade. They were encouraged to 'intensify' production, planting the new varieties in dense monocultures, and to buy the fertilisers and herbicides now needed to handle the ecosystem services once provided by the leguminous shade trees for free. Without those trees, there was nothing working with soil bacteria to fix nitrogen in the soil and nothing to shade out the weeds, which grew like crazy. The new intensified system was called sun coffee.

As part of the course they were teaching in Costa Rica, Perfecto and Vandermeer did some field experiments with their students to look at how the intensification of coffee agriculture affected different organisms. In one study, Perfecto looked at how the loss of shade affected ground-dwelling ants and was surprised to find that the change vastly reduced their numbers. She followed up on that short experiment with a proposal to the National Science Foundation (NSF) for a big grant to study the impact of coffee intensification on biodiversity. Year after year, her proposal was turned down because the people reviewing it complained that she didn't include any forest controls — she wasn't comparing the biodiversity of the farms to the forest. The implication was that the biodiversity of the farms by themselves was of no interest.

'I was kind of pissed off,' Perfecto told me. 'Because my proposal was about biodiversity, the NSF kept sending it for review to conservation biologists who said I had to have a forest control, even though my questions were all about agriculture.'

The couple couldn't find good research sites in Costa Rica where the farms bordered forests. Then, a friend told them about Finca Irlanda, a 300-acre organic coffee plantation two hours from Tapachula. One of several plantations in Mexico started by German immigrants in the nineteenth century, Finca Irlanda claims that it was the first farm in the world to export organic coffee, back in 1930. Ethnobiologists had been doing some studies there, plus a biologist colleague had done research comparing spider density in Finca Irlanda and Finca Hamburgo, the intensified sun-coffee plantation next door. So Perfecto and Vandermeer reached out to Walter Peters, the third-generation owner of Finca Irlanda. A bit of a naturalist himself, Peters welcomed research that would tell him more about the plantation's biodiversity. For the first few years, the couple and their students lived and dined with Peters — he has a large house with a huge wraparound porch informed with posters of bird species seen in the area, both native and migratory, and a series of fenced areas to the side, where his collection of adopted wild animals live. When Perfecto and Vandermeer got more research funding and students, they moved operations to an abandoned, vine-draped building in a forested section of the plantation, where they and their students have worked, eaten, and slept ever since.

Perfecto's first big grant came from the National Science Foundation and supported her collaboration with Russ Greenberg of the Smithsonian Migratory Bird Center and Guillermo Ibarra, from ECOSUR (El Colegio de la Frontera Sur) in Tapachula, both of whom were studying the effects of Latin American deforestation and coffee intensification on birds. People in the eastern United States had noticed that songbird populations were declining right around the same time that the modernisation of coffee farming was expanding; birdwatchers and conservationists realised that changes to the more traditional coffee agroecosystems could be a cause. The Smithsonian had already been investigating how the loss of tree cover and biodiversity adversely affected birds, but what they wanted Perfecto to do in the course of her five-year study was quite different. They were working on a Bird Friendly certification for coffee growers, and they wanted her to find out if having birds on the plantations benefitted the farmers.

'You can't tell a farmer to leave all the trees or plant new ones just because that benefits the birds, because not all farmers care about that,' Perfecto told me. 'We wanted to find out if birds contributed to the productivity and sustainability of the farm.'

To conduct the study, Perfecto and her students distributed netting that prevented birds from getting at the coffee trees in one section of the plantation. They quickly observed that the population of spiders increased when the birds were kept away. Simple logic suggested that the increase of spiders would improve coffee productivity, as the spiders are generalist predators that eat many things, including the herbivore insects that eat the coffee plants. But the interactions among living things are never simple, and they found, paradoxically, that there were both more spiders and more herbivore insects inside the netting. They eventually figured out that the spiders — especially the web spinners — not only eat herbivores, but also catch and eat the parasitic wasps that eat the herbivore population. When the birds reduce the population of spiders, that tends to increase the population of wasps — and *that* pushes down the abundance of herbivore pests.

If Perfecto grew up with an appreciation for a healthy matrix, buttressed later by her studies, Vandermeer came to that appreciation through a more theoretical route. He fell in love with the tropics in the late 1960s while he was working on his PhD and became what he calls a 'conviction-filled conservationist' eager to establish preserves. He soon signed up as a postdoc with Richard Levins, an ex–tropical farmer turned ecologist and epidemiologist. Vandermeer says Levins was the most brilliant person he ever met, and Levins later called Vandermeer the 'outstanding terrestrial ecologist of our generation'. These congenial colleagues travelled in the late 1960s to Puerto Rico. There, Vandermeer started to rethink the value of preserving little patches of wilderness.

While they were in Puerto Rico, Levins and Vandermeer observed that the island had very low biodiversity — for instance, of its native land-dwelling mammals, only bats remain. 'Puerto Rico is an island that's bigger than any patch of wilderness you could preserve in the tropical world, yet it has very low biodiversity,' Vandermeer told me. 'Generally, you have low biodiversity on islands. Dick and I talked about that endlessly.'

As far back as the nineteenth century, scientists had known that species often go extinct on islands, especially small ones. If they are close enough to other islands where that same species lives, allowing migration between the islands and recolonisation, they can regain that lost diversity. E.O. Wilson and ecologist Robert MacArthur published a book about this in 1967, arguing that island biodiversity is maintained by recolonisation from other islands — which happens all the time — balancing out local extinctions, which also happen all the time. A few years later, Levins developed the concept of metapopulations: most species exist in a scattering of subpopulations distributed in habitat patches throughout the greater landscape or on islands or in some other way separated from each other. Any of these subpopulations might die out, but as long as there is migration and recolonisation from one of the other groups, the species as a whole will not go extinct. Without migration, though, local extinctions will accrue until there is a global extinction of that species.

The tropics themselves are now fragmented landscapes dotted with patches of wild nature, separated by land dominated by human activity, usually farming. Applying Levins's concept of metapopulations to what he saw in the tropics, Vandermeer realised that protecting these wild patches would never suffice to prevent species from going globally extinct. Even wildlife corridors that connect the patches aren't a solution, one study indicates, because predators often figure out that prey are being funnelled into the corridor and lie in wait for them. Vandermeer understood that the only way to ensure steady migration and recolonisation is to maintain a healthy matrix that provides some cover and food — and forgoes poisons — allowing species to move around freely.

Without that healthy matrix, species can't migrate safely

from one patch to another. Vandermeer points to a 2017 study of sixty-three German preserves, where researchers detected steep declines of flying insects over twenty-seven years. 'It's such a no-brainer that the agriculture going on outside those preserves is what's killing them,' he says. 'Insects don't respect political or legal boundaries. When they fly out into an agricultural matrix heavily laced with pesticides, they get killed.'

But Finca Irlanda provides the kind of agricultural matrix where the rest of nature thrives, because supporting nature was always a goal for its management. Walter Peters's passion for birds led him to plant eight species of legume trees — most other shade plantations have just one — plus nearly a hundred other kinds of trees. That's about half of what the virgin forest would have at that elevation, but far more than other coffee plantations. The result is a colourful, ear-splitting, sky-disrupting abundance of birds. Peters told me he spots some forty species every morning. Sightings have included rare, threatened birds like the highland guan, which needs fairly pristine forest to survive and was likely migrating from one forest patch to another, enjoying the fruit spilling from one of Peters's trees.

Finca Irlanda has been a research paradise for Perfecto and Vandermeer, whose work over the last few decades has examined the ecological complexity residing in such agriculture as well as the benefits that complexity offers the farmer. Their research takes place within a forty-five-hectare parcel within the plantation's 300 hectares. Over the years, they have determined that the cranky tree-dwelling *Azteca* ant is the keystone species — the species whose presence maintains the ecosystem — of the plantation (and likely of the native forest, too, although the forest is not the subject of their research). As we set out for a research site, they show me a rough map of their forty-five hectares dotted with black circles to mark the presence of shade trees with *Azteca* nests inside, each with ten to twenty queens, and with queenless satellite nests in the coffee plants. Most of the nest cluster sites are named for someone or something that gibes with their political interests — there is a Rojava, named for an autonomous area of Kurds in northern Syria, and a Marichuy, named for an Indigenous woman who was an independent candidate for president in Mexico — or for one of their former students. Not all their students research the *Azteca* — some have studied lizards, mice, or plants — but much of the research looks at the intricate web of relationships anchored by the *Azteca* and its impact on the landscape.

Not that the *Azteca* is found everywhere within the plantation — that's one of the mysteries the team's research has illuminated. As we arrive at one of the sites — called Azteca Crossing, because the ants hurry along the upraised roots of the sope as if they're on an elevated highway — Perfecto turns over some of the leathery leaves of the coffee bushes surrounding the sope to show me how the ants make a living. What looks like sticky white rice along the midrib of the leaf is actually another insect called the coffee green scale, a pest on the plantation that sucks sap from the coffee plants and reduces their yield. The ants farm these insects, protecting them from parasitic wasps and other predators and feasting on the sweet honeydew the scales secrete.

Again, simple logic might suggest that the plantation would be well served by some sort of spray to control the ants, since they both bite the occasional farmworker and protect the scale pest. Instead, the complexity of relationships on the plantation suggests that the best way to protect the coffee from the scales is to stand back and allow what Vandermeer and Perfecto call autonomous pest control to continue — because a third player in this ecological drama is also lurking nearby, waiting to play its part.

When the right kind of phorid flies swarm the *Azteca*, one of them will quickly lay eggs on an ant's head. The doomed ant then emits yet another pheromone that tells the rest of the ants to save themselves by freezing in place. That effectively hides them from the rest of the flies, which only see a panoply of images and can't actually find the ants unless they're moving.

When the ant releases the 'freeze!' pheromone, the third player — a lady beetle called *Azya orbigera* — arrives. The beetle can detect that chemical alert, but only if she is a female nearing the end of her pregnancy. She seizes the opportunity to hustle past the normally vicious but now frozen ants to lay her eggs near the scale insects.

'The beetles are myrmecophilic,' Vandermeer says, 'meaning they love to be with ants.' Because now, what was formerly an unapproachable ant-patrolled refuge for scales becomes a nursery for the beetle larvae. The larvae don't move much. They don't need to: they munch away on the scales, while being protected from the ants by a waxy covering that sticks to the ants' mandibles if they try to attack them. Three different parasitic wasps try to get at the beetle larvae, too, but the ants fight them off, as they can't distinguish the parasites that attack the beetles from parasites that attack their scale livestock — in fact, scientists can't raise the beetle in labs because without the ants' vigilance, their larvae are infested with these parasites. When the larvae reach maturity in the scale refuge, the resulting adult beetles fly away.

'These beetles are the main control of scale on the farm,' Perfecto says. 'So, when farmers see these concentrations of *Azteca* and scale, they have to be willing to sustain some damage because they are the refuge and incubator of the beetles.'

The concentrations of ants and scales never get too big, though: of the 7,000 to 11,000 shade trees in their research plot, only about 700 have *Azteca* nests. The scientists have also tracked a number of interactions that limit nest spread. Phorid flies can bind up the ants in so many freezing cycles that they can't forage or continue their other work — and they die out in that site. Even when the *Azteca* manage to assemble a large herd of the scale insects, that abundance attracts not only the *Azya* beetle but also another player to the ecological drama. This fourth player is a fungus called *Lecanicillium lecanii*, which arrives to feed on the scales, wrapping its mycelia around the insects so that each looks as if it wears a white halo. The fungus can eliminate the scales, starving the ants.

The biggest threat to coffee growers is another, far more pernicious fungus: the coffee-leaf rust, first discovered on wild coffee in East Africa in 1861. Spores arrived in Brazil in 1970, likely via stratospheric wind currents, and began to waft up the continent. Farmers were terrified that the rust would ruin their crops, and with good reason: it had famously done so in Sri Lanka, back in the days when it was called Ceylon. The British colonialists had planted a shade-free monoculture of coffee in that island country, which rust completely wiped out by the late nineteenth century. For a while, it seemed that the disease would just cause minor problems in Mexico — its characteristic yellow spots only showed up in Central America in the 1980s — but it finally detonated its full destructive force there in 2012. Some farmers lost up to 90 per cent of their coffee plants; farmworkers lost jobs, small landholders lost farms, and migration to the United States increased.

'If you had been visiting here back then,' Vandermeer tells me, waving at the lush plantation around us, 'you would have just seen sticks.'

Why did it take the fungus nearly forty years to decimate the Central American coffee farms? Some experts suggest that climate change was the underlying cause, but Perfecto and Vandermeer are convinced that the region's steady drift toward deforestation and agricultural intensification was to blame. Government, foreign aid groups, international banks, and industry had begun pushing farmers in that direction even before the rust scare, but they used the threat of a rust epidemic to push even more extreme modernisation. Frantic to protect their land and livelihood, many farmers took exactly what Perfecto and Vandermeer thought were the worst steps possible: they accelerated the landscape carnage by cutting down their shade trees and spraying fungicides.

One thing everyone knew about rust was that it needs moisture to flourish. Spores blow in or pass from one touching leaf to another, and when they land on a leaf they germinate if there is even a pinprick of dew. Then the rust penetrates the leaf through the stomata, where it grows and destroys tissue, finally re-emerging through the stomata to release more spores. Professional agronomists convinced many farmers that the humidity from the shade trees created moist conditions for the rust spores to germinate, but Perfecto and Vandermeer argue that the shade trees actually protect the coffee plants from rust: they provide a canopy that buffers the coffee from the airborne spores. In fact, their removal allows spore-laced winds to blow closer to the ground — right at the level of the coffee plants.

But the heart of their argument is that the sun-coffee prescription for rust — removing the trees and replanting with densely spaced, genetically identical coffee plants that are heavily fertilised and sprayed with insecticides and fungicides — dismantles the complex ecology that naturally suppresses pests. The rust fungus has at least two natural enemies at Finca Irlanda and other shade plantations. One is a fly that eats the rust but is killed by the insecticides. The other is the white halo fungus that arrives to feast on the *Azteca* ants' herd of scale insects. Without the *Azteca* collecting and concentrating the scale — unwittingly preparing the table for the fungus — *Lecanicillium lecanii* would likely not be able to thrive and grow into a fungal army big enough to halt the spread of its cousin, the rust. And without the shade trees, there are no *Azteca* ants.

Even though the predation of the white halo fungus on rust is what Vandermeer, Perfecto, and their students have studied, it appears that it is not the only fungal rust predator. The group collected hole-punch-sized samples from leaves that had white halo fungus glommed onto rust and sent them to a lab run by University of Michigan's Timothy James, who used DNA analysis to determine that there were actually some 100 fungal species in the patches. Along with the white halo fungus, thirteen were from fungal families that act as parasites of other fungi. They were likely all gorging on the rust.

In an intensified system, of course, many of those beneficial fungi are killed by fungicide spray — even as the rust itself is often spared, since much of its life cycle takes place inside the coffee leaf. Unlike many other plantations, Finca Irlanda does not spray and appears to have less rust than many other farms in the region — without the cost and labour and landscape-devastating tactics the intensified operations employ. Walter

Peters and others who farm this way have confidence that healthy ecosystems can handle most biological challengers. As one of Perfecto's former students from Guatemala found, some of these farmers don't even understand the concept of fighting pests.

For her PhD research, Helda Morales decided to study traditional methods of pest control by interviewing highland Maya farmers in Guatemala, then conducting her own experiments using their methods to see if they made biological sense. But her project seemed to stall immediately when she asked the question, 'What are your pest problems?' The farmers uniformly replied that they didn't have pest problems. But Morales then pivoted and asked them to describe the insects they found on their farms, and they came up with long lists, including some known pests of maize and beans. They explained to Morales that they managed their ecosystem so that these insects didn't get so numerous that they became pests. Besides, they said, those insects also have a right to exist. They planted a little extra for them!

Perfecto had a similar experience with Walter Peters years ago. As they were walking through the plantation, she mentioned to him that she had seen caterpillars all over one of the shade trees, munching away on their leaves. He shrugged and told her he wasn't worried: the migratory birds would soon return from the north and take care of them. And within a few weeks, the migratory birds were back and the caterpillars never had the chance to rise to the status of pests.

Farmers like these know their land and its flow of biodiversity in ways that operators of big industrial operations can't. The latter manage the problems of broken ecosystems with the silver-bullet approaches peddled by huge corporations, approaches that presume every farm has the same challenges, because that mindset yields a manageable portfolio of products and bountiful corporate profits. However, not all landscapes and farms are the same. Industrial agriculture's nod to this reality is something called precision agriculture, in which more costly tools — GPS-driven tractors and lasers that level fields — supposedly help farmers achieve higher yields with fewer inputs of water, fertilisers, and other chemicals. But as Perfecto and Vandermeer point out in one of their papers, farmers like Peters have such an intimate relationship with their landscape that they already do precision agriculture, albeit one based on recognising the particularities of their own land and respecting the relationships among the living things there.

'If you maintain the complex interactions within these ecosystems, they autoregulate and prevent giant pest outbreaks,' says Zachary Hajian-Forooshani, a graduate student who studies natural enemies of coffee-leaf rust at Finca Irlanda, as well as at Perfecto and Vandermeer's other research station in Puerto Rico. 'The precautionary principle is to maintain the ecology that is there.'

Hajian-Forooshani and his fellow students gave presentations on their projects in the research station every night. They spoke in both Spanish and English, an accommodation for me as well as for one of the newer students still learning Spanish, and the presentations were attended by both the Mexican supervisor of the station, Gustavo Lopez-Bautista, as well as a few of the other farmworkers and local residents who work with the research team. Each student was questioned after their presentation, then everyone clapped — and each time, the noise set off Walter Peters's peacocks, which sounded like cats calling 'help!' in falsetto. Sometimes the rain pounded with such volume and intensity that I was sure the dirt steps cut into the road up the hillside back to my room would have washed away, and that I'd fall in the mud and crush hundreds of the tiny toads hopping around.

But the mornings were clear while I visited, so I followed people as they did their research, pelting them with questions and trying not to get lost. Thanks to Hajian-Forooshani, I was learning a little bit about mutualisms involving the native plants and insects there. Like the *Cecropia* with its giant, frog-hand leaves: it is one of the few trees without an overcoat of vines, thanks to the *Azteca* that live inside its hollows and eat a special diet of tiny egg-like nodules that the tree produces on its stalk, seemingly just for them, in exchange for protection against herbivores and smothering vines. Like the *Heliconia*, which spark the plantation and forest greenery with what look like balletic configurations of red and orange birds. One of Vandermeer's first grad students studied the interactions between the larvae of two different fly species that live, with other insects, inside *Heliconia* flowers in Costa Rica. Instead of competing with each other for resources, as he originally assumed, the larvae make life easier for each other. The larva of one fly is a chewer, and the larva of the other fly is a sucker. The first chomps down parts of the flower, leaving some to feed bacteria, which the other then hoovers up. Wandering around, I was sure that these intricate and essential interactions were taking place in every pinch of space. It would take a thousand scientists a thousand years to figure it all out, and maybe not even then.

While Hajian-Forooshani was studying interactions that will have direct implications for managing an actual farm, others were doing more esoteric research. One morning, I followed Lauren Schmitt along a path that left the field station and plunged downhill into a section of the plantation that Peters has allowed to return to forest. The path cut through both denser greenery and more extravagant flowers as it headed toward the river. At clearings along the path, we could see the tidy, sunlit, bird-barren rows of Finca Hamburgo on the opposite hillside. Schmitt had seen an ocelot along this path, so I kept my eyes open for it.

Schmitt scrambled around to retrieve clusters of leaves she had sewn together with fishing line and left in the forest. She hadn't made the task of finding them easier for herself by flagging the sites; she and others have found that it's best not to, as the farmworkers notice and sometimes try to help the researchers by hacking away weeds with their machetes and wind up inadvertently botching the work. So, she consulted a notebook and occasionally swooped down on arrangements of leaves that would have escaped my notice.

Long before I arrived at Finca Irlanda, she had carefully gathered 2,600 leaves from coffee plants and two of the most common shade trees on the farm, *Alchornea* and *Inga*, collecting them after they dropped naturally since trees will resorb some of the leaves' nutrients before they let them go. With these particular leaf clusters, she was going to be looking at how the microbiomes of these different leaves interact as they decay. Each leaf hitting the ground comprises three microbiomes — the one inside the leaf, the one on its surface, and the one on the surface of the soil — and she was interested in seeing how those microbiomes interact and converge. Research shows that leaves decompose faster in a polyculture with shade trees like Finca Irlanda than in a monoculture like Finca Hamburgo.

'We frequently talk about the benefits that the shade trees have — how they support other organisms and can sometimes fix nitrogen — but do they increase the decomposition rate?' she said. 'Because that's a good thing, as it creates more nutrient cycling.'

She held up a leaf with some bites taken out of it. Another focus of her research was the impact of insect herbivores on the rate of leaf decomposition. In most cases, she said, these insects aren't trying to eat the leaf itself but are gorging on the bacteria and fungi on the outside of the decomposing leaves. 'Most people who are eating peanut butter on crackers aren't really interested in the cracker, but you still get a bite of it from time to time,' she explained. 'Same way here. Leaves are a lot of lignin and other structural tissues that are not as nutritious as the bacteria and fungi colonising them. But sometimes they accidentally bite the leaf.'

As I left Finca Irlanda on my final day and observed the neighbouring farms with dismay, I started to wonder how quickly wildlife would return to a farm transformed from a monoculture into a polyculture. Fortunately, I knew someone I could ask: Tom Newmark, whom I had met several years ago. He was a founder of The Carbon Underground, a former chair of the Greenpeace Fund USA, and now, I knew from social media, a partner at Finca Luna Nueva in Costa Rica.

Newmark first visited the Luna Nueva farm when he worked for the New Chapter supplement company and needed an organic source of ginger and turmeric for its products. He fell in love with the place, and he and his wife ultimately joined the farm as business partners and part-time residents.

Finca Luna hadn't always been the Eden I saw on social media, he explained. Most of the land had been a botanical horticulture plantation, where crotons and other exotic houseplants were raised in monocultures, doused in chemicals, and shipped to the United States for our desktops. When that business went bankrupt, a farmer named Steve Farrell took it over and turned it into an organic and biodynamic farm, ploughed with oxen. 'It was doing better ecologically, but it was still a monoculture,' Newmark told me. 'We were growing crops in the kind of regular rows that you'd recognise. Still, we thought we were doing everything right.'

Then in 2008, Newmark and Farrell met Tim LaSalle, former CEO of the Rodale Institute, now the cofounder of and a professor at the Center for Regenerative Agriculture and Resilient Systems at California's Chico State University. LaSalle convinced them that monoculture farming, even organic, was not the most ecological way for them to raise food. Inspired, Newmark and Farrell worked with Rodale to launch a tropical farming systems trial looking at the impact of different kinds of farming in their environment, and baseline testing revealed that their soil was shockingly low in carbon. It was an epiphany that led to immediate change. They remained organic, but stopped ploughing, stopped planting in monoculture rows, and pursued what is called syntropic agroforestry. Based on advice from permaculture designers and famed Costa Rican agronomist Rafael Ocampo, all their crops — such as turmeric, black pepper, cinnamon, and cacao — now grow in the nooks and crannies among thousands of native trees, including the more than 100 species of fruit trees that they planted. In a square metre of cropland, there might be twenty other planted species growing along with the native plants that naturally take root. 'Everything we do now is through a carbon lens,' Newmark says. 'Everything is about getting organic matter into the soil. And it's had a profound effect on the wildlife.'

As the beach almond trees matured and bore fruits in 2018,

Newmark and Farrell found both the scarlet macaw and the green macaw visiting the farm, even though locals hadn't seen them in decades. Mammals that haven't been spotted in years have also reappeared. Agoutis, coatimundis, and tayras are now all over the farm, as well the bigger creatures that eat them: jaguarundi, ocelots, jaguars, and pumas. Birders are now flocking to the farm, and they've sighted some 275 bird species.

The farm is still productive, but differently. Cacao is now their main crop, and they also harvest huge quantities of fruit. 'It was vanity to try to force this rainforest environment to produce like a temperate prairie, in monoculture rows,' Newmark says. 'This ecosystem loves cacao trees — they're native. It loves vanilla vine, also native. It loves allspice and nutmeg and mace. We don't have to do much more than trim and harvest them. And just as we vote with our forks, birds vote with their wings. Right now, we're winning the election with the citizens of our ecosystem.'

Hearing Newmark's love for his chosen ecosystem, I thought back to John Vandermeer and his life-altering trip to the tropics in the 1960s. His and Perfecto's lives at Finca Irlanda hardly seemed like work to me, despite the fact that they put in long hours in the field — hot, muddy, mosquito-swarming hours — and long nights back in the research station. It seemed like such a joy-filled effort with fascinating colleagues, although, of course, that might just be an outsider's perspective. Maybe it was a huge hassle hosting all those students. Maybe the lack of creature comforts at the research station gets old. When I was there one morning for breakfast, there was a burst of worried chatter from the kitchen, where the local women prepared breakfast for the team six days a week. The water smelled bad, and it turned out that the neighbouring farm had sprayed some nasty chemical the day before and it had gotten into our water supply. It took a few hours for Gustavo Lopez-Bautista to drive off and find enough bottles of water to get us through the day.

So, when I told Vandermeer on a follow-up phone call that I think of Finca Irlanda as paradise, I quickly acknowledged that I was probably romanticising their life there.

'Oh no, Kristin,' Vandermeer said without hesitation. 'I think of it as paradise, too.'

## Chapter 7

# Healing from Ridgetop to Reef

'It's very difficult to have an orchard here,' Eric Horvath told me, looking momentarily glum at the sight of a whittled stump.

We were standing some twenty feet from the north fork of Oregon's Beaver Creek, in a clearing off a little-used gravel road where an old homesteader's cabin burned to the ground in the 1970s. The fire happened decades before Horvath and his wife, Claire Smith, bought this property in 2000. Bracketed on all sides by the Siuslaw National Forest, some ten miles from the Pacific Ocean, the land was previously owned by a timber company that clear-cut the sixty acres along the steep slope on the other side of the creek, and then planted a dense monoculture of commercially valuable timber. It was a gross simplification of a previously complex landscape, and it happened all over the Pacific Northwest. And it's just one example of how human actions can unwittingly rupture one of the most essential relationships on our planet — that between the land and the water — with tragic repercussions for the living things inhabiting both.

Horvath and Smith live in the coastal town of Newport, and they bought this land — a total of eighty acres from the road to the top of the ridge on the other side of the creek — not to settle here, but to nudge it back into a healthy, pre-European-onslaught landscape. That's good news. The even better news is that they are not unique. Environmentalists who battled government and private industry to protect nature through the last third of the twentieth century — remember the spotted-owl battles? — say that Horvath and Smith and thousands of other landowners around the state represent a new phase for the movement. They are healing private land in an effort to restore clean, clear, cold water to the state's troubled waterways.

'I've been fortunate enough to watch this transition evolve,' says Paul Engelmeyer, a legendary activist, Audubon property manager, and chair of the MidCoast Watersheds Council. Back in the 1970s, he was hired to plant trees in the aftermath of clear-cuts. It seemed like a great life until he realised how dangerously the logging was fracturing the forests and the implications that had for wildlife. He and other activists insisted that the government measure the impact of all that logging on streams, and it turned out that every Oregon stream failed to meet the water quality standard of the Clean Water Act. He demanded that the EPA list the streams as impaired.

'If they're impaired, then we have to bring them back into compliance,' Engelmeyer explains. 'And that's the first step. But we also realised a few decades ago that we don't just need folks to change the rules. We need folks to sit in people's living rooms and explain that they can work on their own land to improve water quality for salmon.'

The MidCoast Watersheds Council was formed in 1994 to work with landowners, government, and non-profits to reverse human damage with projects that allow natural processes to resume, and it now works across nearly one million acres and five major river basins. Sometimes landowners are inspired to donate land that becomes part of a larger preserve. Sometimes they hold on to their land, and the MidCoast Watersheds Council helps them secure the funding, resources, and even physical labour needed to restore it. Horvath and Smith bought this damaged land with no purpose other than to restore some of the ecological functions it had a hundred years ago.

Even though they're trying to rewild the bulk of it, Horvath and Smith had hoped to keep the tiny patch around the crumbling stone chimney slightly more domestic, with a small orchard and a vegetable garden. But the rest of nature had other plans. Bears climbed the apple trees and snapped the branches. Beavers waddled out of the creek to gnaw down the saplings.

So, Horvath and Smith ditched the idea of having an orchard for humans and decided to plant one for the rufous hummingbirds that arrive every spring from Mexico. They had noticed that the hummingbirds were already drawn to the site, as they'd seen the females peck at the grit holding the blackened chimney stones in place — getting a calcium boost for their eggshells, they assumed. When I visited in early February — too late to see the bright-red coho salmon depositing their eggs in the creek and too early to see the fry emerge from the pebbles or to spot the rust-breasted rufous on their return to the Pacific Northwest — Horvath had already planted several red-flowering currants, a native bush that blooms in March. He was planning to plant another fifty of the bushes, which the MidCoast Watersheds Council will provide.

'Will the bears eat the currants?' I asked.

He shrugged, cheerful again. 'I don't really mind what the bears do.'

Horvath pointed to the steep slope across the creek. The previous owner of the property planted thousands of Douglas firs there, so closely spaced that hardly anything else could grow. Such a dense monoculture is nothing like a naturally growing forest, where many species of tree stretch at their own pace toward the sunlight. When they die or are uprooted by storms or landslides — such deaths are common in a natural forest — they crash to the forest floor and become a gold mine of ecological services to the nature around them. By one estimate, some two-thirds of all wildlife species use broken, dead trees (called snags) or fallen wood for food or habitat. Forest soils are steadily fed and enriched by this woody manna from the overstorey.

Such precious detritus is quickly dragged away from commercial plantations and many other managed forests and parks, but Horvath is eager to have more of it. Ever since he got the land, he's been hiking into the fir-dark mountainside with his chain saw, trying to simulate the patchiness of a natural forest by cutting some of the Doug firs above shoulder height and creating both snags and wood fall. In those open patches, evergreen huckleberry, salmonberry, elderberry, salal, and other shrubs have filled in to create a brush layer that nurtures even more wildlife.

Now, Horvath is conducting that same thinning process with the trees he planted when he and Smith bought the land. The timber company planted some 30,000 trees on the denuded sixty acres, but it left some open spaces. Horvath had eagerly filled the gaps with 5,000 Oregon natives, including western red cedar, hemlock, and spruce. 'We planted 200 trees each day, which was a lot of effort, especially on that steep ground,' he told me. 'When we started, I'd kiss each one and pat it and plant it nicely. But toward the end, I'd just open the ground and slam it in.'

Those trees are now crowding each other, and it's hard work to thin them. He's grateful when he finds that the bears have been roaming among his young trees, ripping away the bark for snacks and leaving them to die. A little break for his chain-saw arm!

Horvath and Smith are birders who lead birding tours around the world, but though they are delighted by the diversity of birds in their eighty acres — the hummingbirds, the thrushes, the warblers, the wrens, the raptors, the woodpeckers, the grouse that whirrs on an alder tree just out of sight, broadcasting his excellence to nearby females — they are restoring this landscape for salmon. Horvath is thrilled when he sees forty salmon in a day in his stretch of the creek, but there used to be many thousands in these waters — enough to keep five nearby canneries busy.

Overfishing itself didn't decimate the population — in fact, according to Mark Kurlansky's *Salmon: A Fish, the Earth, and the History of Their Common Fate*, Indigenous populations used to catch a lot of fish for consumption as well as trade. Early European explorers reported that Indigenous people along the Columbia River used to pull some eighteen million salmon from the river every season. But they didn't destroy the ecological health of the overall landscape that supports the salmon life cycle, as did the white people who displaced them. The timber industry hacked trees from the landscapes, beginning with the big old trees growing along rivers and streams that shaded the water and kept it cool enough for the salmon. Salmon, it turns out, are especially sensitive to warmer-than-average waters even before they hatch — one study showed that Chinook salmon hatch earlier and emerge less developed in warmer water. Ranchers turned lowland forests into fields, opening creeks and streams to the daily glare. Trappers removed millions of beavers, those useful rodents that add ponds, oxbows, and other complexity to waterways and slow the water down so that it can penetrate the surrounding landscape. Industrialising Americans, it seems, have always been eager to force a division between water and earth: straightening waterways, building dikes or even cement walls alongside them, and ensuring that water rushes toward the seas. This creates a flow that is too fast for many living things, curtails the movement of creatures that need both land and water to survive, and parches the surrounding landscape.

There are still some big and old trees on Horvath and Smith's land that the clear-cutters either missed or left behind to satisfy a state requirement to spare a few trees for wildlife. Such trees can be a boon for avian fussbudgets like the locally treasured marbled murrelet, a bird that — like the salmon — requires intact habitat on both land and sea. Called the enigma of the Pacific because so much of its twenty years is hidden from human eyes, it doesn't build a nest but finds a branch high in an old tree heavily felted with moss, shapes a depression there, then lays a single large egg. The murrelet parents take turns incubating and then feeding the chick, jetting back to the open ocean in low light — it's an eight-ounce bird that can fly sixty miles per hour; birders say it's easier to hear its calls than to see it — where it plunges 200 feet down and swims with its wings to catch fish. It was the last North American bird to have its nest officially discovered, in 1974, and Paul Engelmeyer was the first to photograph an Oregon nest in 1991. Birders gather in pre-dawn groups near the coast, hoping to see or hear the murrelets as they leave their chick — by now, the depression in the moss has a corona of white poop — and streak off to work.

Despite their unique value to the murrelets and many other wildlife, Horvath loves it when these giants topple, hoping they will crash all the way down the slope and into the creek. Many of us have the mistaken idea that healthy rivers and streams shouldn't be cluttered with fallen trees, but that's just because we are prejudiced to think the point of rivers is for them to be navigable by our boats. But as Kurlansky says, 'Nature had generally intended salmon rivers to be full of fallen trees.' They encourage natural complexity: water pushes around them making a swirl of currents and sometimes forcing a bend in the stream, gravel and stones gather against the fallen trees, and deep and quiet pools form behind them. Young salmon can hide under logs and escape predators, but still dart to the surface of those quiet pools for insects that would have quickly darted downstream if not for the logjam.

One of the early changes Horvath and Smith made to their land was to have Forest Service helicopters drop seventy logs into their half-mile stretch of Beaver Creek, an intervention made possible by a partnership among the MidCoast Watersheds Council, the Forest Service, and Oregon Fish and Wildlife. These were full-grown trees thinned from stands on Forest Service land, each with a length that is three times the width of the creek with the branches intact, dropped to make crisscrossed piles of two or three so that storms or floods couldn't dislodge them.

As Horvath led me down to the water, he stopped to admire one of the red cedars he'd planted twenty years earlier, stalwart in the sunshine let through by a dying alder tree. 'Just look how fat he is!'

Down at the creek's edge, the water gleamed of sunlight and trees. Modest white rapids twenty feet upstream made the sound of polite applause, but the water in front of us was slower, its surface variously wrinkled and curved with motion. It was so clear that I felt I could count every small stone at the bottom if only I had the time. Horvath pointed out the altered patterns in the bottom of the creek bed where the female salmon created redds (from a Scottish word for making an area clean and tidy) for their eggs and recounted the drama of it. How the female returns from the open seas to the fresh water where she herself hatched; how she finds her preferred spot and beats away the small stones; how she releases her eggs as the male quivers nearby and squirts them with his milt; how the eggs and milt meet in the water and tumble down together and stick to the stones; how she flips on her side and beats the creek bottom again, so vigorously that her tail loses colour, and covers the eggs with more stones. 'The fertilised eggs live in the interstitial gaps between the gravel,' Horvath said. 'They're not being smothered with sediment, but are being washed by this lovely clear water.'

His is not a completely intact landscape, he explained: there is the matter of the road, twenty feet behind us, that bleeds dust and silt through the year. Some of it is screened by a protective border of shrubs and trees between the road and the creek, but some makes its way into the water. It clouds the creek, then settles into the gaps between the pebbles and can smother the eggs before they hatch. The hazard is far less now than when the slope on the other side was aggressively logged, unleashing a dry tide of silt and sawdust into the watershed. But, of course, that kind of activity still goes on in the area — Horvath told me that just downstream, another landowner has harvested timber from the hillsides and gouged the land in the process.

As he led me along this modest tour of his land — we didn't even attempt the other side of the creek — Horvath toted a bundle of mossy sticks in his arms. Too short for shillelaghs, too green for firewood, these were pieces of Hooker's willow for the task that would absorb him after I left. All along the creek, he drives these willow staves into the soft ground, where they readily root and sprout. He hopes they will survive the beavers — they love gnawing on willows — and thrive on the creek bank, where they will unfurl a mighty root system below to keep banks from eroding, and spread branches and leaves above to shade the water.

The clear, cool water resulting from Smith and Horvath's work will not only keep sediment from smothering the eggs, but work like theirs could help protect young salmon and other fish from an affliction linked to warm, muddy waters called whirling disease that's found in other rivers (the disease has not yet been found in Oregon's coastal rivers or creeks like theirs). The disease is caused by *Myxobolus cerebralis*, a member of a family of microscopic parasitic animals called myxozoans that are related to jellyfish and coral. There are some 3,000 known myxozoans. Most don't seem to cause problems for their host fish, even though they inhabit tissues throughout their bodies.

'They're interesting to scientists because their relationships with the fish are very specific,' says Jerri Bartholomew, director of the John L. Fryer Aquatic Animal Health Laboratory at Oregon State University. 'Some infect muscles, others infect gills, some infect scale pockets, and so on.' Bartholomew was part of a team that recently discovered that a fascinating salmon-dwelling myxozoan called *Henneguya salminicola* so effectively steals nutrients from its host that it doesn't even need oxygen — the first such animal known to scientists. Still, it doesn't seem to hurt the salmon.

But a *Myxobolus cerebralis* infection can be devastating, at least to salmon and trout in the United States. It is an invasive parasite that piggybacked entry to the United States in the 1950s, along with brown trout imported from Europe to stock our rivers, and its spores spread through many watersheds. *M. cerebralis* requires two hosts to afflict the salmon. First, the spore needs to infect a half-inch worm in the stream called *Tubifex tubifex*, which is common around the world. The worm serves as a sort of changing room for the parasite, which then departs its wormy host in a form that allows it to drift along until it attaches to a salmon or trout and gets inside the skin. There, it travels to the cartilage in the fish's head and begins to feed. If the numbers of the parasite reach a certain level there, the fish loses the ability to swim properly — it will swim erratically, sometimes in circles — and often dies.

Watersheds disrupted by human activity like logging provide the perfect environment for a whirling disease outbreak. Streams that have lost their cooling fringe of trees have warmer water, which both allows the parasites to reproduce more quickly and reduces the salmon's ability to resist them. And creeks and rivers with silty bottoms — unlike the crystalline water in Horvath and Smith's stretch of Beaver Creek — are the worms' preferred habitat.

In early 2020, Paul Engelmeyer was supposed to take me on an exhaustive tour of the MidCoast Watersheds Council's other projects to undo the human damage to landscapes that impedes the salmon's epic journey from creeks to ocean and back again (in another generation), but, of course, the coronavirus pandemic and the resulting shutdown knocked that off the schedule. By midsummer, I was finally able to tag along with a university team that was filming Engelmeyer's standard tour for students, who usually attend in person rather than online. Five of us rattled along gravel roads, raising a nearly impenetrable cloud of dust, from one site to another in separate cars. There, we stood at a distance as Engelmeyer juggled the charts and maps and comparative photos he's used in innumerable talks.

Salmon need more than cool, clean water in creeks for their eggs and fry. The adults have to have clear passages to those creeks, but are often blocked from some of Oregon's most pristine rivers by human hindrances as large as a dam or as small as a poorly designed culvert. Engelmeyer discussed the latter hazard and efforts to address it during a tour stop at Alsea Bay. The tide was out, and he pointed to a dotted line in the mudflats. 'We just missed a herd of elk,' he said, then gestured at the sky: 'Cedar waxwing!' All the rest of us whirled around, as if he had just shouted the name of a visiting celebrity.

Just out of sight, behind a scrim of vegetation, was Starr Creek, where a culvert used to block salmon in the bay from moving upstream to spawn. Thanks to a culvert-removal project, people are now spotting salmon and their redds in Starr Creek for the first time in twenty-five years. An even more remarkable project took place north of where we stood, when a partnership among organisations and individuals replaced a culvert under Forest Road 1790 that for sixty-two years had blocked salmon trying to spawn in North Creek, a tributary to Drift Creek that flows into the Siletz River.

Only a few years after the old culvert was installed in 1957, people realised that it created two problems for salmon. The culvert jutted so high over North Creek on the downstream side that salmon couldn't make the leap over its lip to access the sixteen miles of pristine forest stream above it. The culvert also blocked the normal, downstream flow of small boulders, gravel, and wood that — before the road was built — used to settle farther downstream, creating great habitat for both spawning and rearing. Instead, as the water poured out of the old culvert, it scoured the creek bottom to bedrock — not good for spawning.

People tried to correct the problem with various devices to help the salmon reach those sixteen miles of pristine creek, but all of them failed. Finally, the MidCoast Watersheds Council undertook four years of grant-writing and fundraising to assemble the nearly $900,000 needed to remove the old culvert and install a new one that restricts neither fish going upstream nor debris washing downstream. The effort itself spawned a massive collaboration among funders, including the US Forest Service, the US Fish and Wildlife Service, the Oregon Watershed Enhancement Board, the Oregon Department of Transportation, the Oregon Department of Fish and Wildlife Fish Passage Program, Trout Unlimited, the National Fish and Wildlife Foundation, and the Native Fish Society, which conducted a crowdfunding campaign. Installed in the summer of 2019, the project was already successful by 2020. Even though salmon typically return to the waters where they themselves spawned, there is nonetheless a natural stray rate — 'a genetic failsafe', Engelmeyer says — and some 10–20 per cent of a returning population will recolonise areas beyond their own spawning grounds. Those strays and their offspring are now spawning beyond the new culvert, swimming alongside other species like the Pacific lamprey, which was also excluded from the upper North Creek.

Not only will the salmon benefit from their newly extended spawning grounds, but the forest alongside North Creek will benefit from the renewed flow of marine nitrogen provided by the dying salmon. 'Minks, dippers, coyotes, river otters, and so many others will be pulling those salmon carcasses out of the stream and into the woods,' Engelmeyer said. 'Things want to heal so much.'

Our little tour stopped next at Lint Slough, an estuary that abuts Alsea Bay, where the removal of another seemingly minor human contraption has made a huge difference for salmon. The slough had once been the site of a failed salmon-rearing facility, and its abandoned concrete structures impeded both the tidal flow of salt water from the bay and the fresh water from Lint Creek. The slough never got enough water from either source and remained too shallow and warm for salmon. After years of planning and fundraising, the MidCoast Watersheds Council managed to remove the structures and restore the estuary. Now, it's perfect rearing habitat for young salmon smolts before they head out to the open sea.

Only fairly recently has it become clear to scientists how important estuaries are to the salmon life cycle. In fact, there was once a suspicion that estuaries might be a sort of Bermuda Triangle that contributed to salmon decline, because their more open inland waters could make the salmon more susceptible to both predation and competition for food with other species. As fisheries biologist Dan Bottom notes, estuaries were once considered a 'bottleneck to production', given the rather industrial mindset driving salmon research back then. 'When we did start studying estuaries in the 1970s,' Bottom says, 'it was all about what is the carrying capacity of an estuary. How many fish can you stuff in there?' There were even goofy gambits in the 1980s to boost salmon survival by catching smolts and carrying them out to sea by barge, thereby keeping them away from the supposed perils of those estuaries altogether.

But of course, salmon evolved to make use of the entire water landscape, and research in the early 1970s by biologist Paul Reimers demonstrated that. He studied a small salmon population in Oregon's Sixes River and found that there were five different ways that salmon reared within that watershed. Some salmon emerged from their redds and swam immediately to the ocean. Some stayed in the creeks for part of a year before heading to the ocean, some for an entire year. Some left the freshwater creeks and swam to the brackish estuaries in the spring, others in the fall, and continued to mature there before they left for the ocean. While the salmon that migrated to the estuary in the fall weren't the largest group, they had the strongest survival rate.

Scientists call this assortment of rearing behaviours — or life-history patterns — a portfolio, and compare the diversity of the salmon portfolio to the diverse investment portfolios financial gurus recommend to build wealth. Just as smart investors spread their money in both stocks and bonds, in both established companies and high-risk tech startups, salmon hedge their survival by 'spreading risk in time and place', as Bottom says. Within a healthy population, there will be some salmon following each of these different patterns (and doubtlessly others that scientists haven't noticed yet). However, the overall landscape has to be intact for them to be able to pursue these diverse life-history patterns, which means that environmentalists also need a diverse portfolio of restoration strategies to heal the disrupted relationship between land and water.

On a seemingly pristine atoll in French Polynesia, scientists are testing a very different land-based restoration to rescue yet another maritime creature. Teti'aroa came to the attention of most Western eyes in 1960, when actor Marlon Brando arrived in nearby Tahiti and Moorea to play the doomed Fletcher Christian — the rebellious first mate to Captain Bligh — in a remake of the classic *Mutiny on the Bounty*, shot in the locale where the historic struggle actually unfolded. Brando not only fell in love with his co-star Tarita Teriipaia, a local woman who had previously worked as a dishwasher and waitress in a local hotel, but he also fell for Teti'aroa, which had once been a private retreat for Tahitian royalty. He married Teriipaia, bought the atoll in 1966, and spent much of the next twenty-five years there.

'If the mermaids can't sing for me here, Christ, they never will,' Brando said. He planned an ecolodge on Teti'aroa — perhaps its most famous recent guest was Barack Obama, who came for a month after the end of his presidency to work on his book — and his estate established the Teti'aroa Society, which welcomes scientists to help preserve the atoll's beauty and ecological integrity.

Teti'aroa, Tahiti, and the rest of the Society Islands

archipelago are bits of land that are the most distant from the world's landmasses — so distant that they were the last to be settled by humans, around 1000 CE, by seafarers from Southeast Asia. Teti'aroa itself is not like an island in the way many of us usually imagine one. Instead, it is a necklace of twelve small, flat islets called motu that circle a bright-turquoise lagoon, which is all that's left of the caldera of an ancient volcano. In NASA photos, the lagoon looks like a jagged piece of turquoise dropped on the lapis lazuli of the South Pacific Ocean. Around the ring of motu — which are the remains of ancient coral reefs that surrounded the volcano — is another, larger circle of reefs.

The first Europeans to stumble upon Teti'aroa may have been three deserters from the historic HMS *Bounty*, in 1789, but they were only a transient presence. Bizarrely, the atoll passed into non-Polynesian hands in 1904, when the family of King Pōmare V gave it to a Canadian dentist who was serving as Britain's consul to Tahiti and tended the bad teeth of the grateful king and his fellow royals. The dentist started a coconut palm plantation on the atoll, and Brando later bought it from the dentist's heirs.

The dentist found Teti'aroa to be distressingly infested with rats, and imported cats to try to wipe them out, but the rats have their own fascinating and persistent pedigree on the island. The Polynesians who settled the Society Islands centuries ago packed in some of the accoutrements of their culture, and one of these was the Polynesian or Pacific rat. The

Polynesians brought them intentionally: they were pets, food, and occasionally couture, as the pelts were sometimes sewn together to make cloaks for the elite. When Europeans landed in the islands, the sailors unintentionally unleashed another alien species — the black rat, which infested many European ships. Both rats reproduced rapidly and took their toll on the native wildlife, exterminating some species. As the rats continue to strip the islands of certain seabirds today, scientists suspect they are disrupting essential biological relationships — and that the damage reaches all the way out to the beautiful coral reefs that ring the island.

Those of us who grew up with little home aquariums dotted with chunks of dead coral as a backdrop for our goldfish have no idea how fascinating coral really is. A few years ago, I visited Oregon State University's Department of Integrative Biology, and inside a dim laboratory, a glass tank full of living corals glowed like the sky during the grand finale of a fireworks display. The tank didn't have a research function; these varied examples were odd bits left over from one of the biologists' research projects and would never actually share the same locale in real life. They were dazzling in their diversity: chocolate fans measled with phosphorescent green, whirligig yellow branches, green pincushions, red cacti, and more. One looked like a frilly pink sombrero.

But as different as they appeared, they had a common feature. 'All of these corals are symbiotic,' said Virginia Weis, then the department chair and now a professor on the faculty, peering into the tank with me. 'Every single one of them has algae, and without that interaction they would bleach and die.'

Corals are tiny animals from the phylum Cnidaria, which includes other marine invertebrates like sea anemones and jellyfish. Most members of this group feed themselves by dangling tentacles in the water and waiting for prey to drift by for a sting and a kill. But millions of years ago, corals and single-celled photosynthetic algae got hitched. The corals eat but don't digest the algae — like tiny plants, but without roots or leaves — which instead live inside the corals' soft bodies and feed them some of the carbon sugars they create during photosynthesis. In turn, the corals not only give the algae a safe home, but also feed the algae their own nitrogen waste. This is such a perfect symbiosis, Weis told me, that many corals don't even bother catching prey.

Corals draw researchers not only because of this intriguing partnership with algae, but also because they are the backbone of important underwater ecosystems. This backbone is the result of steady craftsmanship by the coral animals, most of which are only a half inch in diameter, as they build layer after layer of external skeleton from carbon dissolved in the seawater. Colonies comprising these individual skeletal dwellings form what are very much like massive underwater cities. The individual corals share their structures with fungi, sponges, and molluscs, which move into the bottom 'floors' of the skeleton as the corals keep building out and up, and many thousands of other species live on or around the corals.

Because of this wild diversity, coral reefs are referred to as the 'rainforests of the sea', occupying just 2 per cent of the ocean floor but offering food and habitat to 25 per cent of marine species. More than a million species of plants and animals are associated with coral reefs, making them one of the most diverse ecosystems on the planet.

There are probably millions of mutualisms at work in the coral reef ecosystem, ways in which the corals benefit other marine plants and animals along with ways in which those marine plants and animals help the corals and each other. Scientists understand that one way many of the larger fish help corals is by feeding on the seaweed that grows around the reefs. Corals need clear water that lets in enough sunlight for their photosynthetic algal partners — that's why they're found in relatively shallow, clear waters near land — and overgrowth by seaweed can shade the water as well as crowd out new coral establishment on the reef. Parrotfish and other large plant eaters rescue the corals from this dim fate.

'The plains of Africa have grazers like antelopes, wildebeests, and zebras that mow down the grass constantly — they're like the lawnmowers of the savannah,' says marine microbiologist Rebecca Vega Thurber, one of the researchers studying the connection between the rat infestation on Teti'aroa and the coral reefs. 'Parrotfish are the equivalent of those big ungulates in the ocean. Their job is to eat all the plants around the reefs.'

Vega Thurber spent her childhood summers in the Dominican Republic, where her Dominican father, a physician and an amateur marine biologist, taught her to love the sea. He maintained such an extensive collection of carefully displayed, labelled shells throughout their home that her friends thought she lived in a museum. 'I knew how to snorkel almost before I could walk,' she says. 'I always knew I was going to be a marine biologist.'

She thought she would go on to study charismatic ocean megafauna like sharks, but when she got to graduate school she was instead drawn to the exciting new research into the tiniest things in the ocean and how they interact with larger creatures. As a postdoc, she got a National Science Foundation grant to study the relationship between corals and the bacteria that live in a protective layer of mucus over the reefs — coral snot, as many researchers call it — which protects the corals from particles that fall from the ocean surface. Scientists had known since the 1940s that the bacteria were there, but assumed they were pathogens until a biologist named Kim Ritchie cultured some of the bacteria in the 1990s and found that they produce chemicals that defend the corals against disease. By the time Vega Thurber began her work, scientists knew the bacteria were defenders and that they produce protective chemicals and form a barrier to keep out invaders.

Since then, Vega Thurber and other scientists have found that the bacteria do more than defend the corals. 'The algal symbiont provides the coral with sugar, but the bacteria provide amino acids, vitamins, and all sorts of other important compounds to the coral,' she says. 'Animals are really boring metabolically. We can't make vitamins on our own and have to get things like that from food. But bacteria make all those necessary compounds for themselves and their symbionts.'

Coral snot includes not only bacteria, but also fungi, viruses, archaea, and other tiny organisms that find food and shelter there. Vega Thurber has also been studying the resident viruses since the early 2000s, but it is still unclear if and how they are helping the corals. They always infect the corals, but throughout nature, viral infections can provide benefits in a roundabout way. An example for humans, she points out, is that the herpes virus offers us certain protections against other infections. 'People with some types of herpes are less likely to get listeria from food poisoning,' she says. 'Basically, the virus tricks your cells into making surface proteins so the listeria can't get in, because they want their home protected. We think something like that may be true for corals, too. We see a lot of viruses in and around the coral.'

The coral animals themselves are mobile only once in their lives. In their tiny larval stage, they float in the water searching for the best spot on the reef to anchor themselves for life. Many unknown factors drive their decision. Some researchers think one may be chemical signals from bacteria on the reef that help the larvae and may later become part of the coral's highly diverse microbiome.

Another factor seems to be noise. A lab study by coral reef ecologist Amy Apprill from the Woods Hole Oceanographic Institution showed that larval settlement was twice as high at noisy sites than at quiet ones. Not that the corals have ears, Apprill explains, but they are likely drawn to agitation in the water caused by sound waves. But why seek noisy reefs?

'We think they're drawn to the low-frequency noises of a healthy fish community,' says Apprill, who studies coral reefs in Cuba's Jardines de la Reina archipelago. 'Even if there are predators in that community, the corals are likely benefitting from it in many ways that we still don't understand.' One, of course, is that a healthy fish community includes the lawnmowers that keep the sunlight streaming in. But another may be that all those fish coming from other parts of the ocean are bringing diverse microorganisms into the reef environment. These microorganisms live in the mucus that covers the fishes' bodies — yes, fish snot. Apprill thinks it's possible that the larvae are drawn to the sounds of a healthy fish community because having thousands of fish around increases the opportunities for a larva to encounter diverse microbes as it fastens to the reef and begins to select microbes from the water column for its own microbiome.

Despite their protective microbiome, corals face many dangers, both global and local. A major threat comes from chemicals that humans unleash on landscapes, which stream into the water from rivers and underground seeps. Nitrogen in agricultural fertiliser — and even in human waste, from all the nitrogen-fertilised produce we eat — is a huge problem. This human-made source of nitrogen disrupts the coral microbiome, reduces the number of beneficial bacteria, and, according to a study by Vega Thurber and her colleagues, spurs the growth of a bacterial parasite that overwhelms the microbiome, steals energy, disrupts growth, and makes corals more susceptible to disease.

Pollution and other stressors accumulate to enact even greater damage on a reef. For instance, fish bites are another common and natural stressor on the reef: some bite the corals on purpose; some bite them accidentally as they try to grab something else. Usually, the corals quickly recover. But when they are stressed by pollution, they are more likely to die from fish nips — perhaps, Vega Thurber surmises, because their microbiome is so compromised that it can't fight off the pathogens living inside the fish's mouth.

Some stressors cause the corals' algal partner to release chemicals that are toxic to the corals. When that happens, the corals eject the algae. The corals will then 'bleach', and vividly coloured colonies turn into ghostly white landscapes as the corals struggle to survive without their main food source. A change in salinity can do it — Vega Thurber was once doing research south of Australia's Great Barrier Reef when the salinity dropped because of a very low tide combined with heavy rain; within two days, the entire reef they were working on turned white. Bleaching has been in the news since the early 1980s, but the bad news on this front has grown ever more frequent and unnerving as climate change raises the temperature of the seas and all that newly warmed water triggers bleaching.

Recently, parts of the bleached reefs have recovered. According to Andréa Grottoli, director of the Coral Bleaching Research Coordination Network at Ohio State University, a colony of corals will lose some of its algae during a bleaching, but not all. 'They usually have one million algal cells per centimetre squared of the surface area of a coral colony,' she tells me. 'It's a lot, but when they bleach, they go down to 100,000 — orders of magnitude less. Research shows they can grow those back up again.' But now that bleaching related to climate change occurs so frequently — the Great Barrier Reef experienced an unprecedented back-to-back bleaching in 2016 and 2017 — scientists fear that we could lose the coral reefs entirely.

One strategy for helping the reefs survive is to try to eliminate local stressors, even as we continue to do all we can to halt and reverse global climate change. These other stressors are like pre-existing conditions, which — as we have seen with the coronavirus pandemic — can make some individuals more lethally susceptible to a threat that others can survive. For the coral reefs surrounding Teti'aroa, one of the pre-existing conditions may be the island rats that have vanquished the local seabird population, including some ten to twelve species. For that reason, Vega Thurber and her research partner, marine biologist Deron Burkepile, are studying how a conservation group's 2020 effort to exterminate the island's rats will impact the coral reef.

While Vega Thurber grew up steeped in ocean science,

Burkepile grew up in Mississippi — as he jokes, 'a noted hotbed for coral reef ecologists'. But he had a cousin who was a scuba diver and he wanted to be just like him, so he convinced his parents to let him become scuba certified. 'I grew up diving all over north Mississippi and Alabama in some of the nastiest lakes and rock quarries you've ever seen,' he says. 'Once I put my face in the ocean, I knew this was what I wanted to do the rest of my life.'

Unlike Vega Thurber, Burkepile has spent his career studying big animals — first, the big animals of the African savannah, then the big animals of the seas. He studies how the loss of the bigger animals affects the entire ecosystem, extending all the way down to the microbial life at the foundation. For instance, as he and Vega Thurber wrote in a paper called 'The Long Arm of Species Loss', overfishing that eliminates plant-eating fish in the reef ecosystem causes disruption all the way down to the coral microbiome.

The connection between island birds and coral health may not be as obvious, but the two have a connection that goes back thousands and thousands of years — cemented by the humble medium of poop. When the birds were living on the island, they would fly out to fish over the reefs and then return to rest, nest, and raise their young on the motu. As they and their babies poop, they concentrate nitrogen-rich marine nutrients on the island. Those nutrients not only nourish the plants and animals on the island, but they also seep back out to the reef whenever it rains. 'The seabirds are big nutrient conveyer belts,' says Burkepile. 'That big pulse of nitrogen is critical for the reef's health.'

How is it that nitrogen from synthetic fertiliser and other human sources is bad for the reefs, but natural bird-poop nitrogen is good for them? Nitrogen from the birds arrives as ammonium, urea, or uric acid, Burkepile explains, and these are forms of nitrogen that the corals have encountered during millions of years of evolution. They are used to having other creatures urinate or defecate on them! Not only do these forms not damage the reef, but the coral has enzymes that pluck that nitrogen out of the water before their algal symbiont gets it and allows them to parcel it out to the algae to pay them for their contribution of carbon sugar. It makes the algae work harder for the coral, possibly helping the coral become strong enough to resist bleaching and recover more quickly. Conversely, the coral does not have the enzymes to take up the synthetic nitrogen from fertiliser, and the algal partner snatches it up — and thus loses its incentive to feed the coral. External pressures like this can disrupt a partnership and turn a helpful symbiont into a parasite.

With the rats gone — they will be poisoned — Burkepile figures it may take five to ten years for birds to repopulate the island. Before and after the eradication, he and Vega Thurber will be taking samples from the corals, the algae, the seawater and fresh water on the island, and the soil. They will study changes in the area's biogeochemistry — the chemical changes wrought by living organisms and especially by the microbial community. The microbes will be the first responders to the changes in chemistry wrought by the removal of the rats and the return of the birds. 'It's always the microbes that are doing the fundamental chemistry in the soil and water, and often in the animals,' Vega Thurber says. And of course, they will be gauging the health of the reef.

She and Burkepile hope their new Teti'aroa research will highlight the connection between land and water and make people more aware of conservation efforts that can heal both. Scientific research is often weirdly siloed because of funding, they say, but these two ecosystems are intricately linked into what Burkepile calls super-ecosystems. Anyplace where land and water touch — which is just about everywhere, when you consider that landscapes far from the seas contain and are shaped by watersheds — is a super-ecosystem. We humans have the choice to be symbionts with the other creatures in these super-ecosystems. Too often, we have unwittingly — and sometimes with full knowledge — chosen to be parasites.

## Chapter 8

# Living in Verdant Cities

The shrubs and grasses in this Bend, Oregon, garden echo the sere July landscape outside the city. Silver and gold foliage mixed with some occasional bright-green curve along the garden paths, popped here and there with colour from high-desert natives like Munro's globemallow (orange) and pineleaf penstemon (more orange) and sulphur buckwheat (yellow) and flax (blue). Bees climb in and out of the flowers, heedless of the shadow I cast when I bend to look at them. When I straighten up to admire the mountains in the distance — Mount Bachelor and the Three Sisters, still swirled in snow — I walk into a cloud of gnats furiously circling empty space for reasons known only to them. I would have once grumbled about gnats in my face, but given the plunging populations of insects around the world, I'm glad that this garden is bountiful with insects as well as trees and shrubs and flowers and mountain views. I almost apologise to the gnat I have to dig out of the corner of my eye.

It's easy to forget that I'm downtown, standing on top of a roof, two storeys up.

The CEO of the insurance company Moda Health

searched for a year for someone who could create this 15,000-foot garden comprising only native plants on the roof of their new headquarters, and ecologist and nurseryman Rick Martinson eagerly took on the job. Martinson and his partner, Karen Theodore, harvest seeds and cuttings from different microclimates and ecosystems in the high-desert West — some ecosystems can be as tiny as the community of plants growing inside cattle hoof prints, which can differ dramatically from those growing outside the prints — and propagate 170,000 plants from 233 species each year at Bend's Wintercreek Restoration and Nursery. Martinson picks from that collection to design and install landscapes for customers who want plant communities that know how to thrive in the harsh, if beautiful, high desert of central Oregon — and not just thrive, but also support the local birds, insects, microbes, and other living things that populate the area's greater ecosystem.

Martinson is often hired to design a landscape around a building scheduled to be built. He'll spend as much as a year collecting plants and soil from the site before ground is broken, hoping to later reintegrate both the local plants and soil microorganisms into his new creation. They will be the seeds of the site's former ecosystem — its legacy and memories — which will help restore it after the devastation of construction. 'When a building is developed, the soil is dug up and compacted, there's fill brought in, and there are all sorts of chemical and biological changes to the site,' Martinson says. 'The sun exposure changes, the wind patterns shift. I worked around Mount Saint Helens after it blew. The devastation was amazing, but it wasn't as radical as what happens in an urban environment when they put in a building. From an ecological perspective, the site becomes toast.'

Martinson had a different set of challenges for the Moda rooftop garden. He was creating an ecosystem where none had existed before, in a city with scant rain and frosts that can strike even in the summertime — and on top of a building, to boot, where the wind and glare are extreme and the soils are shallow. But Martinson managed to select forty-five species of native plants that would succeed in these conditions. 'You can find a native plant that's adapted to just about any condition,' Martinson says. 'I've had urban foresters ask if there's a native that will work in a four-by-four sidewalk cut-out. And yes, there are plants adapted to that condition.'

He filled the rooftop beds with soil that was half compost, because it would retain moisture and would make a good home for the native fungi and soil bacteria that he introduced along with the plants. As he does in all his designs, he arranged plants in groupings that replicate what he's observed out in the countryside when he's collecting specimens: certain plants seem to like to grow near other plants. 'You observe that there are distinct relationships in nature, even if you don't really understand them,' Martinson says. 'You pick up on the subtleties if you look long enough.'

Twelve years after installation, the Moda garden is flourishing. A few of the thousands of plants Martinson bedded there have died, but others have stretched out to fill the empty spaces. A handful of native species have arrived by wind or bird and made themselves at home. And other people around the country have picked up on the idea: Moda's was the country's first green roof comprising all native species, but now there are others. This is one of the great environmental trends gathering momentum around the world. In ways that are subtle — most green roofs are on private buildings, thus out of sight — and bold, people around the world are doing the very important work of greening our cities, softening one hard surface after another with life, inviting the rest of nature to live among us, and trying to replicate the conditions that allow it to flourish.

I knew I wanted to close this book with a chapter on cities, since that's where most of us live. More than 80 per cent of Americans live in cities, and more than half of the rest of the world's people are likewise urban. Compare that to 1960, when twice as many people lived in the countryside as in the cities. I was riding my horse every day after school back then, cut-off jeans and bare legs pressed against her dusty back, on land that is probably now hardened with streets and houses. Development continues voraciously, with the United States losing some 4,000 acres every day — three acres per minute — to new streets and homes and businesses, each acre undergoing its own ecological apocalypse at the hands of the surveyors and earthmovers.

We tend to think of cities as places that are lifeless, except for us, our pet animals, and the rodents and cockroaches that thrive in our presence. Artists have variously referred to cities as 'concrete jungles' or 'asphalt jungles', images of grim, grey confinement far from the rest of nature.

But such images aren't really accurate, as cities have parks, street trees, and even odd undeveloped patches, as well as the lawns and gardens of homeowners, apartment complexes, and businesses — and many creatures nest and shelter in all that urban flora. The new city-greening trend urges us to not only add layers of nature on hard surfaces, like roofs and walls, but also to value and protect the fragments of nature that are already among us — to let them expand, if possible, and grow a little wilder. For instance, a grassroots movement in London including 'cyclists, scientists, tree climbers, teachers, students, pensioners, unemployed, underemployed, doctors, swimmers, gardeners, artists, walkers, kayakers, activists, wildlife watchers, politicians, children, parents, and grandparents' mounted a campaign to declare London a National Park City, a recognition that even though at least 8.6 million people live there, a whopping 47 per cent of the city is physically green, including ancient woodlands and meadows, parks and newly created wetlands, as well as private gardens. Added together, those many green areas comprise a huge park, and the idea is to extend the stewardship principles now applied to national parks to this vast patchwork of nature within the city. London mayor Sadiq Khan signed a charter proclaiming this new status in 2019. Other cities are likely to follow.

We need a vibrant connection to the rest of nature to be our best selves — more on that later — and given the global biodiversity crisis, the rest of nature needs our cities for habitat, food, and respite. Studies show that cities are already providing that: in one, researchers found a higher diversity of native bees in several American cities than in nearby rural areas.

To get a better understanding of this worldwide city-greening trend, I planned to go to its temporary epicentre: a conference in Paris convened by The Nature of Cities (TNOC). This is an international network of visionaries dedicated to cities that are 'resilient, sustainable, livable, and just' founded in 2012 by David Maddox, a New Yorker with a varied background in science, conservation, music, and theatre. What he loved about theatre was that each show was a collaborative event, with each person part of something they couldn't have created on their own. 'Cities are also fundamentally collaborative events,' Maddox told me. 'People collaborate all day long to make them better. My idea with TNOC was to pull together different points of view — scientists of various kinds, designers, architects, artists, activists — to have conversations about the kinds of cities we want. Together, we might be able to create something we could not individually attempt by ourselves.'

He was happy to have me lurk and listen at the conference. It was going to be a wonderful week of doing that, followed by an equally wonderful week of travel with a friend.

Alas, I broke my right leg a few weeks before the conference and went without my friend, dumping the plans for the extra week of travel. Despite the delights of Paris, I can confirm that it's miserable to be there alone, booted, and on crutches. I missed the field trips showcasing France's urban-greening standouts, the part of the conference that I was most looking forward to. So I didn't get to see the Boulogne-Billancourt science school with the forest on its roof or the community gardens of northern Paris, or the arts activities flourishing in an abandoned, overgrown former railway, or the 'linear' park along another abandoned railway, or the Abbé-Pierre–Grands-Moulins wetland, or the wasteland areas that have been transformed into eco-districts, or the two urban parks that host 140 bird species. On the day of the field trips, I stayed in my hotel room binge-watching *Dead to Me* on my laptop.

But aside from the disappointment of not being able to see any of Paris other than my hotel, the café across the street, and one small part of the Sorbonne, it was an amazing conference.

Tim Beatley, a University of Virginia professor of sustainable communities, ran one of the workshops I attended. He has dubbed these consciously greening urban areas 'biophilic cities', a tip of the hat to biologist E.O. Wilson's concept of biophilia (love of life): the notion that humans have an innate desire to live in proximity to the rest of nature because we evolved with the rest of nature and thrive in its presence. In 2013, Beatley formed an international coalition called Biophilic Cities, in which urban areas aspiring to this greater greenness share experience, inspiration, and tools. 'Our goal is a truly nature-immersive urban environment,' Beatley told me. 'We want to reimagine the city as a place of nature and part of an interconnected ecosystem, with nature at the very centre of design and planning. At their best, these biophilic cities protect and incorporate nature at every level and every scale, from rooftop to region.'

The poster child of the Biophilic Cities group is the island city-state of Singapore, represented at the conference by Lena Chan. She's a senior director of the International Biodiversity Conservation Division of the city's National Parks Board (NParks), the lead agency responsible for greenery, biodiversity conservation, and wildlife and animal health, welfare, and management. NParks manages the city's four nature preserves: more than 350 parks and gardens; 225 miles of park connectors, including walking, jogging, and biking paths surrounded by greenery, so that humans and other creatures can move between the bigger parks without leaving green space; ninety-five miles of Nature Ways, which are roads with multitiered plantings to emulate a tropical rainforest; a million managed trees; 295 acres of skyrise greenery; 1,600 community gardens tended by more than 40,000 gardening enthusiasts; and the Singapore Botanic Gardens. Landsat images show that even though the city's population grew by two million people from 1986 to 2007, the percentage of urban greenery increased, from 36 per cent to 47 per cent. Singapore has also been successful in increasing the numbers of some native species, including dragonflies, butterflies, otters, and tropical birds called hornbills. The otters have a Facebook page called OtterWatch.

'A sky park or green wall isn't a replacement for the ground-level nature we want to experience, but we don't want to consume more nature with urban sprawl,' Beatley says. 'Singapore has helped us imagine what is possible in the vertical spaces of very dense cities, where there isn't a lot of space at ground level.'

Singapore began this work more than fifty years ago when former prime minister Lee Kuan Yew proposed that the city should reconceive itself as a 'garden city', which he felt would make it more competitive for foreign investment. Singapore's leaders understood what the rest of our warming, storm-wracked world is just catching up to: that these greening initiatives make cities not only more beautiful, but also safer and more liveable. To address today's key challenges — biodiversity loss and climate change — Singapore is evolving itself into a 'City in Nature', which will seek and employ even more nature-based solutions for climate, ecological, and social resilience.

Cities are typically hotter than the surrounding countryside; despite our quite understandable fascination with hurricanes and floods, heat is actually the leading weather-related cause of death in the United States. A dense urban tree canopy offers protection from this heat, creating shaded surfaces that can be a whopping twenty to forty-five degrees Fahrenheit (11–25 °C) cooler than unshaded surfaces. Other greenery also reduces urban heat; even a green roof high above the pavement extends its cooling effect to the sidewalk below. Urban greenery cuts back on air pollution, too, as leaves filter out the fine particulates that come from dust, burning fossil fuels, or the smoke from forest fires. Plants also remove carbon dioxide from the atmosphere and feed some of it to their partner microorganisms in the soil, where a percentage of the carbon remains sequestered.

Land planted with trees and other vegetation also absorbs stormwater and decreases urban flooding, reducing the expense of adding more costly pipes and facilities to protect the built environment from excess water. Even small pockets of greenery deployed in swales and rain gardens suck down massive amounts of stormwater, allowing it to dissipate gradually into the ground, rather than overwhelming city streets, drains, and stormwater facilities. Cities are better protected against megastorms by an abundance of these 'soft' green surfaces than by hard infrastructure. Susannah Drake, an architect and landscape architect in New York and founder of DLANDstudio, argues that when we rely on engineered solutions to manage water — metal pipes and concrete channels and so on — we develop a false sense of security that nature can be controlled. 'We manage water systems to within an inch of their life,' she tells me. 'But the engineering no longer matches the force of natural systems. There's much more water falling all at one time and few unpaved surfaces to absorb it.'

Green roofs like the one Martinson installed in Bend —

and the 208 green roofs in Singapore — also help individual buildings stay cooler in summer, warmer in winter, absorb their own stormwater, and protect the roof membrane. While the initial costs of green roofs are higher than conventional ones, a green roof will last longer and save the owner, some $200,000 (in 2016 dollars) over its lifetime. Green roofs can be an astounding eighty degrees cooler on their surface than conventional ones, so most of the savings stem from a reduced need for air conditioning.

Cities swathed in greenery also keep citizens healthier. Research keeps showing what most of us instinctively know: that proximity to nature bolsters and heals us. Back in 1984, scientists discovered that merely seeing trees through a window helped surgical patients recover more quickly and need fewer painkillers. Continuing research has shown that the effect is more than psychological. For instance, while the sight of trees and plants and flowers brightens our spirits, we also know now that plants release chemicals — which we inhale or absorb through our skin — that boost our immune systems.

People around the world are picking up on the Japanese practice of forest bathing, or shinrin-yoku, but we don't have to head for the wilderness to glean the benefits from the plant world's inadvertent ministrations — we can experience them in well-greened cities, too. And it's not just a boon for our individual health; science shows that exposure to plants also has a profound effect on our social health. One study shows that people living in areas with more greenery report 'lower levels of fear, fewer incivilities, and less aggressive and violent behaviour ... the greener a building's surroundings were, the fewer crimes reported.' Another found that the more street trees a neighbourhood had, the fewer prescriptions for antidepressants were filled. Still another study found that public-housing residents 'living in buildings without nearby trees and grass reported more procrastination in facing their major issues and assessed their issues as more severe, less soluble, and more longstanding than did their counterparts living in greener surroundings'.

After years of working to support and expand nature in Singapore, Lena Chan has come up with her own set of down-to-earth metrics. She says that a city becomes more biophilic when the area of green cover and tree canopy increases every year; when more natural and human-created habitats are enhanced and restored with native species; when the known number of native species escalates due to discovery of new species and rediscoveries of species thought to be extinct; when the participation rate of citizen scientists expands; and when more than 50 per cent of the residents can name and recognise at least ten native plants, birds, and butterflies.

Chan's final point about the importance of humans having their heart in this movement — you can't help but care about other species if you recognise them and know their names — was made over and over at the conference. Honestly, I hadn't thought of this before. I assumed that if urban areas created the conditions for the rest of nature to thrive, humans would eventually notice and be jazzed by it. Instead, many of the visionaries at the conference talked about taking active steps to jolt public interest and build a constituency for nature in their midst. Facts and dire predictions about the bleak state we'd be in without the rest of nature don't necessarily motivate, but a bunch of cute otters on Facebook or some great storytelling or fabulous art might help. One of the first presenters, in fact, showed how a group in New York has brilliantly incorporated the arts into their campaign to have a Bronx waterway called Tibbetts Brook 'daylighted' — a term that refers to the various ways people around the world are uncovering and unleashing creeks and other waterways buried by decades of urban development.

Most cities were built on top of thriving ecosystems near rivers, lakes, or oceans, lush confluences of land and water that have long attracted communities of living things. In his marvellous book *Mannahatta: A Natural History of New York City*, ecologist Eric W. Sanderson recounts just how richly biodiverse the island once was. 'If Mannahatta [the name given to the island by the Lenape people who lived there when Henry Hudson's Dutch and English sailors arrived in 1609] existed today it would be a national park — it would be the crowning glory of American national parks,' Sanderson writes. 'Mannahatta had more ecological communities per acre than Yellowstone, more native plant species per acre than Yosemite, and more birds than the Great Smoky Mountains National Park. Mannahatta housed wolves, black bears, mountain lions, beavers, mink, and river otters; whales, porpoises, seals, and the occasional sea turtle visited its harbor.'

Sanderson goes on listing the wonders of the island's flora and fauna, and later notes that 'Mannahatta was copiously well-watered, with over twenty ponds, sixty-six miles of streams, and, it has been estimated, three hundred springs.' The nearby Bronx — and most other cities — was also well-watered with ponds, creeks, streams, brooks, and springs that squiggled and squirted through the greater watershed and later became inconvenient to the burgeoning settlement of humans. Tibbetts Brook was one of these. Water from 2,500 acres drain into the brook, which back then meandered toward the Harlem River. But in the early twentieth century, the last mile of the brook was diverted into the city sewers. Now, up to five million gallons of fresh water from Tibbetts passes daily through the Wards Island water treatment plant. Every time there's a heavy rain, Tibbetts' fresh water overflows its underground confines and floods streets. The extra pulse of stormwater from Tibbetts also joins the flow of sewage into the treatment plant, and the combination overflows and dumps waste directly into the Harlem River.

An artist and green activist named Mary Miss took notice. She had founded an organisation ten years earlier called the City as Living Laboratory to enlist artists in making environmental concerns more gripping to their fellow citizens, and she organised walks and community meetings with artists and scientists along the eighteen-mile corridor of Broadway, from the tip of Manhattan to the top of the Bronx. She heard from locals about the flooding and the sewage in the river. She found out that when you walk along parts of Broadway in the Bronx, Tibbetts Brook courses down below your feet — you can even hear it. 'There's a beautiful old brick sewer underneath and there are four to five million gallons of water running through every day that doesn't need to be there,' Miss says. 'And that's on a dry day!'

Community groups came together to develop a plan. They proposed uncovering one section of the brook and directing its flow into a mile of abandoned railroad line, instead of the treatment plant. The railroad land could then be redeveloped as a long narrow park, with walking trails and a bike path alongside the resurrected brook. Tibbetts would once again interact with air and sunlight and the earth along its banks, creating new ecosystems and nurturing biodiversity as it heads toward the Harlem River.

When Miss showed slides of the work to daylight Tibbetts Brook, I almost gasped with excitement. There was a beautiful rendering of the old, historic Tibbetts meander from its headwaters in Yonkers to its terminus at the river, done in blue ink by artist Bob Braine — obviously an image of the waterway, but at the same time, reminding viewers how similarly and vitally the human body is meandered by our blood and veins. Miss's City as a Living Laboratory put together a series of neighbourhood events called Finding Tibbetts, which included a mobile wetlands exhibit. The most popular spot at these events was Braine's table, where he offered to paint his Estuary Tattoos — his image of the old meander — on people's bodies. Hundreds of people of all ages and backgrounds lined up for a blue tattoo, even though it took Braine nearly a half hour to do each one. As he painted, he showed them the old meander painted on top of a contemporary map and talked about what the brook used to be and how the daylighting would return at least part of it to the land's surface.

Braine got the idea for the tattoo from the landscape patterns he sees in satellite images and thought people might be interested in having him mark their bodies with shapes from the land where they live. 'The estuaries outside the body and your own inner ocean are a real connection — not just a metaphor, but real,' he says. 'The tattoos are a way for them to connect their corporeal bodies to the landscape. They got that and they wanted that.' At least two people have gotten permanent tattoos of the historic Tibbetts meander.

Shortly after Mary Miss talked about the Tibbetts Brook outreach, Toni Anderson took the podium and flashed slides that likewise made me want to shout with excitement. She had juxtaposed a map of the annual migration of monarch butterflies from their winter grounds in Michoacán, Mexico, to Chicago with a map of the historic migration of African Americans from the South into Chicago's Bronzeville neighbourhood. This brilliant pairing of one species' struggle and survival with the African American migration — and beauty! — is just one of the tools she uses to engage Bronzeville youth in environmental issues.

Anderson grew up on the South Side of Chicago. Back then — she's now in her early fifties — local African Americans still had connections to rural family down south, and they'd send the kids down every summer once school let out. There, the younger generation would reconnect with people who were still growing and hunting their own food, who knew enough about receiving from nature to want to protect it. 'Slavery was based on our ability to cultivate the land,' she told me after the conference. 'The trauma of that aside, our relationship with the land is in our blood memory. We can wake it up.'

Anderson's mother's family had lost their land down south, but her mother wanted to make sure Anderson spent some time in nature. She found her daughter a scholarship to a residential summer camp run by Chicago Catholic Charities, and she attended from the age of seven through fifteen. 'It was a complete surrender to nature,' Anderson says. 'I knew how to canoe, swim, ride a horse, pitch a tent, make a fire and cook an entire meal on it. These are not the skill sets a young Black girl on the South Side of Chicago grows up with. And when I came home, I suddenly realised how many trees were on my block. I started seeing all the nature that had been invisible to me before.'

Anderson knew most of the neighbourhood youth weren't building that kind of relationship with nature — and that they needed to, with global warming and other environmental crises not only coming their way but disproportionately affecting them and other people of colour around the world. So she started talking to science educators and community leaders about filling gaps in their environmental education and figuring out how to make the science relevant to them. She formed a group called the Chicago Sacred Keepers Sustainability Lab in 2012, which she describes as a group of 'passionate tree huggers, mad scientists, stargazers, activists, and culturalists' with the mission of teaching youth to inherit the Earth.

The group began with a Saturday school as well as activities that engage the community. For instance, Sacred Keepers Sustainability Lab organises a beach clean-up every Earth Day at Chicago's Margaret T. Burroughs Beach. This is not just another dirty stretch of lakefront, but the site of a hideous incident that sparked the 1919 Chicago Race Riot. An African American teenager, Eugene Williams, was swimming with friends and crossed an invisible boundary between the segregated White and Black sides of the beach. A White man on the shore noticed and began throwing stones at Williams, who finally drowned. Years later, the beach was renamed to honour Bronzeville artist and activist Margaret T. Burroughs. So the annual beach clean-up is a double healing, as the group reclaims the beach from garbage and also from its hate-filled past.

Anderson's monarch work began years ago, when she worked with young girls who were wards of the state or recent adoptees whose status was at risk because of behavioural problems. She helped develop a program using the monarch butterfly as a symbol of personal transformation and planted a butterfly garden as horticultural therapy for the girls. She started to wrap in the history of the Bronzeville community, aligning its migration and resilience with that of the monarch. When she started Sacred Keepers, her 'tale of two migrations' caught the attention of the Field Museum and the Nature Conservancy. She and her group were invited to develop a monarch habitat with signage in the Burnham Wildlife Corridor, which is one of the Chicago Park District's designated natural areas.

'The Burnham Wildlife Corridor literally runs along the tracks where the trains from the South brought in the African Americans who would then activate this community,' Anderson says. 'I really wanted to make sure that migration was visible to everyone.'

Anderson now runs a seven-week summer monarch ecology program funded by the US Forest Service and the Nature Conservancy, and up to twenty neighbourhood teenagers take part every year. Their job is to maintain and monitor Sacred Keepers' seven monarch habitats for milkweed host plants and monarch eggs, larvae, and adults. They also help other organisations establish habitats as well as host a community-wide monarch festival.

'They learn all about the ecological aspects of what's happening in the community, and they learn about all things monarch,' Anderson says. 'One of their jobs is to feed information about the sites to our partners, who are looking at what's happening on the ground and telling us what the trends are with monarch population and migration, based on what's happening with climate. So the kids really understand the citizen science and the sacredness of what they're doing.'

There are many ways to gauge the success of this work, and one is to look at its effect on an individual. Anderson tells me that one of the teenaged interns called her not long ago and announced, 'Miss Toni, you blew all my cool points.' He said that he was hanging out with a group of friends and spied a big patch of milkweed. He raced off to count the plants and inspect them for eggs, leaving his incredulous friends behind to ponder this bewildering new obsession. 'He blew his cool, but he stayed there to count seventeen plants,' Anderson says. 'That's exactly what I want, to foster their curiosity and help make the nature around them visible. Once they see it, they can't not see it. They can't not care about it. It becomes part of who they are.'

To create biophilic cities — and we are so accustomed to bleak and dystopian visions of cities that it's hard for us to believe that there can be soft and verdant ones — we need citizens who are that excited about nature. Sometimes it takes a brilliant spark like Braine's Estuary Tattoos or Anderson's tale of two migrations. Sometimes nature in the city puts on such a grand show that we can't help but notice. Throughout the month of September here in Portland, thousands of people gather on the grounds of Chapman Elementary School to watch the evening skies fill with Vaux's swifts, which create a tremulous cloud that swirls and swirls and then suddenly funnels out of the fading blue into an old, decommissioned smokestack. This population of swifts has been roosting there, on its way back to its winter habitat in Central America, since the 1980s. And even though the thousands of humans who attend the nightly show know how it will end, there is always a collective rapturous sigh when the birds vanish. People in Austin turn out in similar numbers, from March to November, to watch North America's largest urban bat colony peel away from the Congress Avenue bridge every evening to forage. When I lived in Cleveland, I was part of an email alert system for people who wanted to watch the annual migration of salamanders from one side of a road to the other to reach their ancestral breeding pools. Every spring, hundreds of people arrived with their flashlights to stand in the rain and watch this silent, scurrying parade.

That eagerness to witness the rest of nature in our midst easily becomes an eagerness to protect it. That can lead to action — and even action on an individual level, multiplied by thousands or millions of urban residents, can be significant.

There are many groups around the world trying to convince urban dwellers to change the way we manage our own small pieces of property so that we create the conditions that allow the rest of nature to flourish among us. To give up the pesticides that reduce the diversity of plants, insects, and fungi; to ditch the fertilisers that disrupt plants' relationships with their microbial partners in the soil and leach toxins into waterways; to maintain a diversity of native plants that support birds, insects, and other animals. To make sure they include some flowering plants for the pollinators. To let our lawns become more like meadow habitat, with clovers and other flowering plants among the grasses, and to let them grow longer than the golf-course length we're used to. To stop compulsively cleaning our properties of biological debris and instead, to let leaves and fallen blooms mulch the ground between plants; to accumulate fallen branches into piles that shelter birds, amphibians, and other small creatures. To make sure wildlife has some access to water, either in a bird bath or in a pond. To reduce the amount of hard surfaces in our yards by removing concrete and asphalt or installing less of it.

But nature needs more than one or two small yards; it needs a bigger canvas if city dwellers hope to help reverse the plummeting survival rates for the world's birds and other species. At the TNOC conference in Paris, I learned about a new British program developed by Earthwatch Europe called Naturehood that encourages whole neighbourhoods to make this shift. Based on a study published in the journal *Biological Conservation*, organisers estimate that if all the private gardens in the United Kingdom became nature-friendly, this would create a wildlife haven of over 430,000 hectares — around 175,000 acres — which is more than four times as large as all the United Kingdom's national nature reserves combined. That would give a considerable boost to struggling species.

The program kicked off in four neighbourhoods, two in Oxford and two in Swindon, and features wildlife-focused events and activities. Organisers work with residents to make urban life easier for five beloved local species: the European hedgehog, the common frog, the house sparrow, the small tortoiseshell butterfly, and the early bumblebee, although, of course, other species will also benefit. The list of recommended garden practices includes those listed above, and special accommodations for the adorable slug-eating hedgehog: participants are encouraged to open up small holes at the bottom of fences or gates so it can conduct its night-time mosey from yard to yard unhindered.

Regenerative farmers have told me that one of the biggest reasons other farmers still raise monoculture crops doused in multiple chemicals surrounded by lifeless bare soil is that it looks neat and tidy, and that they are nervous to have it look otherwise to their greater community of neighbours, landlords, bankers, and so on — it might smack of bad management. Homeowners have similar fears of neighbours and even city officials frowning on what they consider a weedy, untended landscape.

Attitudes are changing, but until they catch up with the ecological imperative of changing the way we farm and garden, homeowners can employ 'cues to care' — a phrase coined by professor of landscape architecture Joan Nassauer. Basically, these are actions showing they are taking care of their garden and understand the community norm.

'I think what offends people is the idea that someone is being sloppy and not doing their job,' says former Naturehood research manager Tristan Pett, who employs one of these cues by mowing a strip around patches of grass that he lets grow long enough to function as habitat. 'People are happier if they understand that what you're doing is intentional and that there's a good reason for it.' One gardener involved with Naturehood decided to make her cue especially vivid: she made a colourful hand-lettered sign for her yard announcing, 'Wildlife Garden — we're helping bees, hedgehogs, butterflies, and more.' Before she put up the sign, a few people had sniped that her garden was messy. After the sign, people were stopping by to take pictures of the garden and sign.

Of course, there are many large and systemic ways that cities must change to make them truly nature-full, and individual citizens or even neighbourhoods can't pull that off alone. So green visionaries must reach out broadly to people with different areas of expertise and roles within the urban environment.

'People who are interested in greening need to have productive, collaborative conversations with people who have other professional interests,' David Maddox says. 'I was in Taipei once at one of these greening meetings, and someone said that we needed to be sure to get our scientific studies into the hands of city planners and into the public dialogue. I looked around and said, "Do you see any city planners or community members here?" We have to keep asking ourselves who is missing from this conversation. Participation is destiny, in many ways: the people who are in the room are the people who will control the outcome.'

As a successful example of this kind of big-tent approach, Maddox points to the process that began in 2015 to pull a wide swathe of New Yorkers together for a series of gatherings to develop nature goals for the city. More than seventy-five organisations have taken part, and the meetings include nature managers, scientists, urban planners, biodiversity conservationists, and advocates for environmental justice. Because of this effort, the city of New York included biodiversity and wildlife conservation in the city's strategic planning for the first time in 2019.

Translating green visions into government policy can have a huge impact. Because of this, the National Wildlife Federation (NWF) is calling upon cities to make three policy changes that would turn tens of thousands of acres of urban landscape into habitat for the iconic wildlife people love — monarch butterflies, bats, and songbirds — as well as the tiny organisms we can't see but that undergird the entire infrastructure of life.

First, NWF urges cities to pass native-plant ordinances so that the landscaping for parks, city buildings, median strips, and other common areas is drawn from the palette of plants native to each area. 'Native plants have evolved over millennia with the insects and other animals that rely on them,' says Patrick Fitzgerald, the organisation's senior director of community wildlife. 'That connection and collaboration make a critical difference. For instance, if you plant a Norway maple in the United States it will attract exactly zero species of Lepidoptera. Plant a red maple, and you attract more than 300.' Fitzgerald says the caterpillars from those butterflies and moths are the essential food for birds like Carolina chickadees, which need up to 9,000 caterpillars to raise a brood of five young.

Second, NWF wants cities to update their weed- and vegetation-control ordinances. Most are based on landscaping protocols that don't match the needs of our biodiversity-challenged world and reflect an older generation's fixation on keeping nature tidy and decorative. Homeowners who want to use native plants to create pollinator meadows or prairie patches — or even just let parts of their lawn grow long and shaggy to provide habitat — sometimes run afoul of those ordinances.

And third, NWF suggests that cities create no-mow zones in parks and other open spaces where native plants and wildlife can flourish. These no-mow zones also reduce city spending for mowing, fertilising, and the use of other chemicals as well as decrease the toxic emissions from those activities. In a YouTube video publicising the decision by Pinellas County, Florida, to establish no-mow zones in less-used sections of the county's parks, park ranger Pam Trass declares that reappearance of some species can be immediate. 'We've had an incredible increase in pollinators,' she says. 'We have a large amount of butterflies, we've had birds that normally you wouldn't see ... and we've seen fireflies over at Wall Springs Park at night. I've never seen fireflies here, and I've lived in Pinellas County for over fifty years.'

These are obvious — although, not necessarily easy, since getting groups of humans to change is never easy — ways to make cities more welcoming to the rest of nature. Green visionaries are also coming out with ideas that are not so obvious and that will require cooperation among not only many groups of people, but also many layers of government. One of the people I met at the Paris conference was Susannah Drake of DLANDstudio. She talks about healing the landscapes and natural systems that development has fractured with green stitches and sutures. She and her former colleague Forbes Lipschitz coined the word 'infra-sutures' to describe this work.

'Ecosystems and watersheds have been fragmented, divided by highways and power lines and other development,' Drake says. 'Our idea was to come up with urban design strategies to knit or stitch together these watersheds, wildlife pathways, and migration zones. But there are a lot of barriers because politics and jurisdictions don't necessarily align with physical geography and ecology.'

Case in point: one of Drake's ideas was to create what she calls a Sponge Park near Brooklyn's 1.8-mile Gowanus Canal, which was the site of early Dutch settlement back when it was a wetland creek and was dredged and straightened to become a busy shipping lane in the 1800s; now, it's an EPA Superfund site that grows more polluted every time it rains as raw sewage overflows the treatment plant and runs into the canal, along with stormwater that rushes along city streets into the waterway bearing chemicals and debris.

The pilot Sponge Park was completed in 2016, and Drake, who has a patent pending on the design and has trademarked the name, envisions many more. The first Sponge Park is an 1,800-square-foot garden composed of a special soil mix and woody plants that can thrive while flooded and remove toxins from water. The site also includes a pedestrian platform across the top to provide access to the waterfront. As water gushes down Brooklyn's Second Avenue during a storm, it flows into Sponge Park, which traps junk like cigarette butts and paper cups as well as chemicals. Every year, Sponge Park filters and cleans some two million gallons of water before it reaches the Gowanus. If taken to scale in New York City, Drake estimates that a series of Sponge Parks could clean some 800 million gallons of stormwater every year.

But this small, brilliant project took eight years to complete. Drake soldiered on for seven years to secure the 200 permits needed to do something like this on the waterfront — 200 permits! — and get funding. The actual construction took only six months.

'It was a prototype, so it was worth tackling to get all the public agencies talking to each other,' Drake says. 'At the time, they weren't communicating well and that was holding back progress on green infrastructure. It was helpful to have then mayor Michael Bloomberg and his leadership team create

PlaNYC [a strategic plan to prepare, among other things, for climate change and enhance the quality of life for all residents]. We could ask agency representatives, "Don't you want to work toward the goal of PlaNYC and your mayor?" Bloomberg and our novel grant-based, community-driven process gave them political cover to take risks.'

Drake is working with others on more big ideas for greening New York. One is figuring how to make use of the city's concrete underbelly, the massive amount of public space beneath the five boroughs' 700 miles of bridges, elevated highways, and transit lines — an area four times the size of Central Park. Sunlight often makes its way underneath these structures, but not water. Drake has created and installed prototypes for capturing highway water from downspouts and cleansing it of chemicals in a planted structure, like the Sponge Park. This approach could support greenery in these utterly grey, semi-subterranean spaces. Greened and sheltered from the elements, they could become venues for music performances, street markets, or even makerspaces for artists and others.

'Often, when you're welding or sculpting you make a lot of noise and dirt,' Drake says. 'But these spaces are already environmentally compromised and aren't next door to where other people are living.'

Perhaps the most ambitious project Drake is involved with is the BQGreen, a massive endeavour to create a 'park out of thin air' in Brooklyn's mostly Latinx South Williamsburg area.

When the Brooklyn-Queens Expressway (BQE) went

through the area in the 1950s, it ruptured South Williamsburg's low-income neighbourhoods with a wide trench along which some 108,000 cars roar and emit roughly 20,000 pounds of pollutants daily. South Williamsburg now has one of the highest asthma rates in the city and is comparatively low in the amount of open space per citizen, ranking forty-sixth out of fifty-one districts in the city by the organisation New Yorkers for Parks — typically, the poorer an area is, the fewer parks, street trees, and overall biodiversity it has.

Drake's design calls for extending a concrete platform over a two-block section of the expressway trench, a feat of engineering that would stitch the neighbourhood back together again — including two small parks now at the edges of the BQE — and add three and a half acres of new park space. The design includes a flower garden, a playground, a baseball diamond, barbecues, grassy and wooded areas, an indoor swimming pool, and a water-play zone. With this single, multimillion-dollar swoop, BQGreen would answer many of the neighbourhood's problems. The platform would block many of the highway pollutants from reaching the neighbourhood, and the parks' trees and other plants would cleanse much of the rest of the pollutants from the air as they simultaneously cool it. The project would swap highway noise for the sounds of people outside, having a good time. And the sounds of birds! Even though BQGreen is an even more ambitious project than New York's wildly successful Highline, it's not an unprecedented design idea: Seattle, Dallas, and

Philadelphia have already created similar green spaces over highways.

I love the artists' rendering on Drake's website of the BQGreen project. Not only is the huge concrete platform that covers the expressway freckled with green and orange and yellow — paint standing in for trees, grass, and flowers — but the colours hopscotch across nearby roofs, too, as if all that new life on BQGreen will keep reaching for new places to take root. As if people will keep finding new places in the city for the rest of nature. That's the way I look at all the hard surfaces in cities, now, after the TNOC conference: every roof, every wall, every underside of an overpass could host a patch of green. Every expanse of road or sidewalk could be broken up, even just a little, to put in a little more vegetation, to allow a little more rain to penetrate the soil instead of filling up the gutters.

And of course, with more vegetation in the city, there will be more animals. More birds, more insects, more of the tiny organisms we can't see without a microscope. Like the moss piglets — aka water bears, aka tardigrades — tiny eight-legged creatures with what looks like a Cheerio for a mouth that live just about everywhere researchers look, including in the tops of city trees. Their superpower: in situations where there isn't enough water or oxygen, they curl into a ball and enter a state of suspended animation for long periods. Someone discovered how long by shaking some of these balls out of a piece of moss that a museum had collected 100 years earlier. With a bit of water, the moss piglets came back to life.

We're not likely to notice a crowd of moss piglets up in our trees, but we do pay attention when large predatory animals show up. In my neighbourhood, an alert streaks through social media within minutes of a coyote sighting. Almost every urban area has large predators, according to urban ecologist Christopher Schell — coyotes, raccoons, and the occasional bobcat in the United States, spotted hyenas and golden jackals in African cities. 'Just knowing these animals are in the area sparks curiosity and conversation,' he says. 'Also, if we have a great diversity of wildlife within cities, we are more likely to have multiple trophic levels and interactions that can positively impact the health of our urban environment. Having a more sustained carnivore guild reduces the likelihood of things like rodents becoming too abundant, which have been in conflict with humans since the dawn of civilisation.'

Schell's love of urban nature was sparked by seeing coyotes in his historically diverse neighbourhood in Altadena, California, while growing up. Now an assistant professor at the University of California, Berkeley, he studies how human-wildlife interactions shape the behaviour and evolution of urban wildlife, and develops strategies to mitigate conflict and promote coexistence. Schell is also interested in the ways in which socioeconomic and racial disparities affect both who has access to nature and who is invited into the ranks of science. He applies that lens to his research lab at Berkeley as well as the Grit City Carnivore Project, which focuses on urban carnivores and their interactions with humans. One of his senior urban ecology colleagues in Madison, Wisconsin, David Drake, runs a similar project focused on coyotes, where the animals are live-trapped and radio-collared in order to learn more about their habits and impact. Drake's program actively solicits community involvement in everything from trapping to data collection. When we spoke, Schell was developing a comparable component for the Grit City Carnivore Project, too.

Many of us tend to admire coyotes' adaptability for urban environments — at least, when we're not freaking out about them eating our cats (although, as feral cats and, to a far lesser extent, pet cats that spend time outside are carnivores that decimate populations of native birds, many people regard coyotes as the hero of that story). Schell explains that many other animals have not only adapted to cities, but have also evolved in them. 'Literally, allele [a type of gene] frequencies in urban populations are changing,' he says. 'For a long time, biologists thought cities were inhospitable places where animals were trapped and couldn't thrive. This is a big change in the narrative.'

Just one example: certain species of lizard have developed longer limbs with bigger toe pads so that they can skitter up slick walls and poles. They easily climb surfaces that their rural cousins find difficult. 'There are growing instances of urban evolution in multiple organisms across the phylogenetic tree of life,' Schell says. 'We think of evolution as something that happens across hundreds of thousands of years. But these animals are like, "No, I got this — give me a couple of generations!"'

Who knows how many other animals have adapted to city life and are living among us without our notice? Portland herpetologist Katie Holzer told me about one that surprised her: the Oregon slender salamander, which scientists thought lived only in mature forests in the Cascade Mountains. Then volunteers pulling up invasive plants in a suburban Portland park found a group of them snuggled into the soil under some ivy. That chance encounter made salamander fans realise they might be living undetected in other city areas, too. 'The biggest site that we've found so far is the cinder-block retaining wall around one of our volunteers' gardens here in Gresham,' Holzer told me when I visited her in this Portland suburb, where she is a watershed scientist for the city. 'It's a rare salamander that we thought only lived in pristine areas, not in the city. And there it was all along, but no one was paying attention.'

She told me this as we walked through the Columbia Slough Regional Water Quality Facility, a multimillion-dollar project to clean stormwater runoff from 965 industrial and commercial acres before the water can empty into, and hopefully not pollute, a local slough. Stormwater sounds innocuous because it starts off as rain. But when rain falls in an urban area, the many hard surfaces there prevent it from sinking into the ground. Instead, the storms create a pulse of water that washes across roofs, roads, industrial sites, and landscaping, gathering pollutants that threaten fish and aquatic organisms at the bottom of the watershed. Roofs yield chemicals to curb moss growth. Roads yield oil and brake-pad dust and chemicals from our tires. Landscaping yields fertilisers and pesticides. Stormwater is a witches' brew of toxic chemicals by the time it gets to the water's edge.

Before I visited Holzer, I would have imagined that the facility to handle this stormwater was a building filled with pipes, filters, and chemicals. But the city of Gresham took the enlightened step of creating a wetland planted with native species to handle the stormwater in this location. After years of planning, in 2007, the city cleared off a field of invasive blackberries that Boeing, an area business, had donated for the effort. Backhoes carved a pattern in the soil designed by engineers, creating a thirteen-acre system of ditches that kink and curve — the original plan reminded me of a diagram of the human intestinal tract. Two huge stormwater pipes drain from the industrial and commercial acres into the engineered wetland, and the water drifts through the ditches and then finally dumps into the slough. The goal was threefold: to clean the water, to create a wildlife habitat, and to provide the public with educational opportunities and a place to connect with nature. Wetland birds like swallows and cedar waxwings found the new habitat right away, as did frogs and salamanders. But while the engineered wetland made the water cleaner, it didn't clean as well as the designers had hoped.

Right around the time Holzer was hired in 2015, someone noticed that the water levels in parts of the wetland were changing. 'No one could figure out what was going on,' Holzer told me. 'The engineers were freaking out!'

It turned out that a family of beavers had moved into the wetland and built dams where the ditches curved, creating ponds. Holzer's department decided not to let the beavers mess up the engineers' plans, so one of her first tasks at the facility was to wade into the wetland and destroy the beaver dams. 'They were an amazing matrix of sticks and mud and grass, all interwoven,' she recalls. 'It was the end of the summer and dry everywhere else, but it was moist inside the dams. Frogs and salamanders were just hanging out there, waiting for fall.'

A few days later, the dams were back. Holzer's department held many fraught meetings trying to figure out what to do. They finally decided to hire a contractor to trap and remove the beavers. But a few weeks later, another family of beavers moved in, because what the engineers had inadvertently done was to create a perfect habitat for them, with lots of moving water, mud, grass, and willows. There were more fraught meetings. Then Holzer suggested that instead of trapping more beavers, they conduct a two-year study of the site with and without the dams and figure out whose system cleaned the water best — the beavers' or the engineers'.

Well, the beavers won, handily. Holzer and her colleagues compared the water coming out of the wetland after fifteen big storms, seven of them with intact beaver dams and eight without. All significant pollutants — mercury, lead, zinc, copper, nitrogen, phosphorus, sediment, and pesticides — were dramatically reduced by the way the beavers changed the way the water flowed. When the dams were in place and the water pooled behind them, the pulse of water during a storm flowed right through the dams — through that matrix of mud, sticks, and grass that Holzer had been ordered to destroy when she first started. As the water passed through, each droplet of water came into contact with millions of soil particles, which grabbed pollutants and held on to them. Some of those pollutants were then broken up by soil microbes.

People have known for a long time that running water through soil is one of the best ways to clean it, Holzer explains. But stormwater-facility engineers have a hard time designing large wetlands that run water through soil structures because they break apart every time a storm sends a big surge of water through. 'We haven't figured out how to do that without having to go out after every storm and fix the breaches, and that's too much work,' Holzer says. 'But that's exactly what beavers do for free — after every storm, they built the dams back up again. It's pretty amazing.'

Holzer says she gets two very different responses when she tells this story to other people who work with stormwater. Some want to do everything they can to keep beavers out, and some hope beavers show up in their facility. She understands the feelings of the first group, even though her heart is with the second. Because working with nature in this way is unpredictable, whereas stormwater management has always been just the opposite. 'We have pipes, boxes, and steel catch basins,' she says. 'We know exactly how big they are, we know how they're going to look next year, and if one breaks we fix it with another one that looks exactly the same. We're used to having control, but there is uncertainty and change with these green facilities.' Even when they're doing a better job cleaning the water, that's hard for some people to accept.

We humans expect much of our own ingenuity and cleverness, but we often fail to consider that many of the answers we seek already exist and are part of the greater creativity — and generosity — of nature. That should give us hope, because we're not alone in our struggle for wholeness. In every ecosystem, we have partners eager to take on the task of restoration, with tools we might not even be able to imagine at their disposal. Many of them are so much tinier than the Gresham beavers that we hardly know they're there, but they are ready to resume their role in maintaining the Earth's ecosystems. When we help them do their job — and often, that just means stopping our damage and getting out of their way — nature rebounds more quickly and bountifully than we imagine. We can trust nature to do this work.

It helps to re-examine our tired, unhelpful red-in-tooth-and-claw assumptions about the world around us. Of course, there is competition, but there is also cooperation and peaceful coexistence and generosity that surprises even hard-eyed scientists.

Not long ago, my partner and I signed up to join a guided group of birdwatchers visiting a protected meadow near Lake Tahoe. All of us whispered our delight that our first sighting — a green-tailed towhee — was sitting right on the sign near the parking lot. We marvelled at the wonder of sparrows throughout the meadow, which seemed so similar from a distance but were delicately distinct when viewed through binoculars, each with a different song or call. We were riveted by the sight of the pygmy nuthatch darting in and out of a tiny hole in a dead tree, feeding its young — so riveted that many of us missed the western meadowlark as it shot by.

At one point, our young guide pointed to the spikey red *Castilleja coccinea* — commonly called scarlet paintbrush, painted cup, or prairie-fire — growing next to the path. She told us that she recently learned that the plant is parasitic and can take water and nutrients from nearby plants. And because I had been working on this book and thinking about the give and take among living things, I wondered aloud if the paintbrush gave anything back to the other plants. Our guide didn't know, but one of the other birdwatchers snickered, 'Why *would* they?'

There was that attitude!

This turned into a personal challenge that involved hours of googling for research and even talking to two scientists — Illinois State University ecologist Victoria Borowicz and her grad student Anna Scheidel — with an interest in this kind of plant. Called hemiparasites, they may not need a host to survive but can grow special structures called haustoria that penetrate nearby plants for a sip. The individual paintbrush that finds a host grows bigger than paintbrushes that don't, and the host dwindles in comparison to others of its kind without a paintbrush glommed on.

Viewed individually, I don't think even I can make a case that the paintbrush is cooperating with its host. But when scientists study plant communities in which paintbrushes mingle with other wildflowers, they often find that their presence enhances the overall landscape. The paintbrushes frequently steal from the dominant plants in the community and thereby create opportunities for less dominant plants to thrive and spread. In that case, they can break up what could be a monoculture of lupines or mule ears — those were the wildflowers I most frequently saw near the paintbrush — and push the community toward greater diversity. The hemiparasites also accumulate more nutrients in their leaves, thanks to the host; when the leaves fall to the ground and decompose, these nutrients are concentrated in the soil in ways that benefit the other plants in the community — ecologists call this the Robin Hood effect. And in what they call the Dryad effect — named after female spirits from Greek mythology who live in trees and create sacred groves as they nurture their own trees as well as the living things around them — this boost to plant diversity attracts more birds and insects to the community, increasing pollination, seed dispersal, and other good things.

One might call these gifts. And maybe, just maybe, scientists will someday discover that the lovely scarlet paintbrush gives a little something special to its host.

# Acknowledgements

I worked on this book for some six years, although my attraction to its ideas has been around for decades. To alchemise that attraction into an actual book took a lot of help. I quake at the idea of trying to thank all the people who helped me do it, afraid I'll miss someone, but here goes.

My agent, Kirsten Neuhaus, was enthusiastic about this project from the very beginning, and I'm so grateful for her enthusiasm and bright, thoughtful support over the years. I am also grateful that Patagonia Books Director Karla Olson was interested in the idea of this book, even when it was just a few blurted words, and patient with the time it took me to turn it into a manuscript. Many thanks also to John Dutton, who cheerfully (or so it seemed!) edited the book, and to fact checker extraordinaire Kate Wheeling — I could kiss her feet for catching my mistakes and misunderstandings. I'm sure some remain; totally mea culpa.

Because I am an English major and have no background in science, I tend to ask the scientists I interview a lot of questions. Then more questions, and more again, and sometimes again. Then I ask them to read what I've written to make sure I've got it right. Thus the scientists in these pages — and even the nonscientists, whom I also ask many, many questions — unwittingly accepted a huge burden of participation when they responded to my initial email, but all were unfailingly gracious and helpful. I'm sure more than one thought, when they saw yet another email from me, 'God, not her again!' But they never responded that way.

I have to stop for a moment and say how lucky I am that I'm the kind of writer who never interviews hateful people — no prevaricating public figures, no scam artists (except one, long ago), no one who makes a living by bruising or belittling or terrifying other people. No one making millions by turning Paradise into parking lots (thank you for that apt alliteration, Joni Mitchell). I get to interview people who are doing remarkable work that may help save all the rest of us from prolonged human error. Everyone in these pages is one of my heroes.

Here are the people I interviewed, including some who are not in these pages because I couldn't figure out how to include their science or stories or perspective: Bret Adee, Athena Aktipis, Toni Anderson, Amy Apprill, Jerri Bartholomew, Tim Beatley, Ian Billick, Robin Boies, Steve Boies, Seth Bordenstein, Victoria Borowicz, Doug Boucher, Nat Bradford, Clyde Bragg, Bob Braine, Mike Bredeson, Judith Bronstein, Anne Buchanan, Deron Burkepile, James (JC) Cahill, Lauren Carley, Jonathan Chadwick, Emily Chan, Lena Chan, Sudeep Chandra, Kelly Clancy (and thanks to my pal Miriam Garcia for sending me Kelly's essay), Pat Coffin, Micaela Colley, Julian Davies, Gopal Dayaneni, Peter Donovan, Sharon Doty, Susannah Drake, Lee Dugatkin, Bruce Dvorak, Melony Edwards, Paul Engelmeyer, Carol Evans, Justin Evertson, Steven

Farrell, John Feldman, Tommy Fenster, Del Ficke, Patrick Fitzgerald, Megan Frederickson, Bill Freese, Jack Gilbert, Lyf Gildersleeve, Kelly Gravuer, Jon Griggs, Andréa Grottoli, Zachary Hajian-Forooshani, Lori Hoagland, Mark Hoban, Katie Holzer, Eric Horvath, Mike Houck, Nicole Hynson, David Inouye, Rebecca Irwin, Chris Jasmine, David Johnson, Hui-Chun Su Johnson, Melissa Jones, Richard Karban, Daniel Karp, Nicole Kirchoff, Stephanie Kivlin, Buz Kloot, David Krakauer, Claire LaCanne, Allen Larocque, Jim Laurie, Kristyn Leach, Connie Lee, Jeremy Lent, Sandra Loesgen, Jonathan Lundgren, Julia Maddison, David Maddox, Rick Martinson, Michael Mazourek, Margaret McFall-Ngai, Katie McMahen, Nicholas Medina, Mary Miss, Todd Mitchell, Tom Mitchell-Olds, Andrew Moeller, Kailen Mooney, Jeff Moore, J. Jeffrey Morris, Frank Morton, John Navazio, Athanasios Nenes, Tom Newmark, Brian Paddock, Ivette Perfecto, Walter Peters, Tristan Pett, Brian Pickles, Mary Ridout, Christina Riehl, Iris Saraeny Rivera-Salinas, Jean Roach, James Rogers, Graham Rook, Marilyn Roossinck, Teresa (Sm'hayetsk) Ryan, Joel Sachs, Eric Sanderson, Anna Scheidel, Christopher Schell, Lauren Schmitt, Jodi Schwarz, Vivek Shandas, Nyssa Silbiger, Suzanne Simard, Gregg Simonds, Agee Smith, Vicki Smith, Tamzen Stringham, Sherm Swanson, Kristine Tompkins, Stan van de Wetering, Laura Van Riper, John Vandermeer, Aaron Varadi, Rebecca Vega Thurber, George Waldbusser, Karen Wang, Ken Weiss, Virginia Weis, Rachel Whitaker, Jeff White, Noah Whiteman, and Jasper Wubs. Thank you for enlightening me.

I also have boundless gratitude for my many writer pals! Thanks to Mary Grimm, Susan Grimm, Mary Norris, and Tricia Springstubb, who were boon companions as I churned through this book, as well as through many other writing projects — we go so far back that there's hardly a word I've written that they haven't lovingly perused. Much appreciation for the support and smart suggestions from other writer pals, too, including Rachel Dickinson, Marina Krakovsky, Charlotte Huff, Jill Adams, Caren Chesler, Cynthia Ramnarace, Sara Cooper, Julie Talbot, Lori Callister, and Celia Wagner. Thanks to Judith Schwartz and Diana Donlon — soil sisters! — for discussing this material with me off and on. Thanks to writer office-mates Lola Milholland and Lee van der Voo — I felt smarter and more productive in your presence. Thanks to Ben Goldfarb for sending me the then unpublished manuscript of his wonderful book *Eager: The Surprising, Secret Life of Beavers and Why They Matter* — we missed each other by days in eastern Nevada, both drawn to the story of the ranchers and their cherished rodents. Thanks to Cat DiStasio for help with some of the early research. Finally, thanks to *Discover* magazine, *Experience Life* magazine, and TakePart media for publishing early versions of some of this material.

Special thanks to Tim Sheils, my first reader, who makes my life sweet in more ways than I can count.

For those readers interested in the sources this book draws from, please visit the book's page on patagonia.com.